JOHN CHANNON

WITH

ROBERT HUDSON

THE PENGUIN
HISTORICAL ATLAS
OF RUSSIA

PENGUIN BOOKS

Published by the Penguin Group
Penguin Books Ltd, 27 Wrights Lane, London W8 5TZ, England
Penguin Books USA Inc., 375 Hudson Street, New York, NY 10014, USA
Penguin Books Australia Ltd, Ringwood, Victoria, Australia
Penguin Books Canada Ltd, 10 Alcorn Avenue, Toronto, Ontario, Canada M4V 3B2
Penguin Books (NZ) Ltd, 182–190 Wairau Road, Auckland 10, New Zealand

Penguin Books Ltd, Registered Offices: Harmondsworth, Middlesex, England

First published 1995
Published simultaneously by Viking
1 3 5 7 9 10 8 6 4 2

ISBN 0–14–0–51326–4

Foreword

Recent years have seen momentous changes in the former Soviet Union. Its demise at the end of 1991 was to witness the emergence of newly independent states, of which Russia is the largest and most powerful. *Glasnost*, under Gorbachev, enabled the study of new archival materials and made possible re-evaluations of the past. Historians in the post-Soviet states are now able to study their own history free from the constraints of Soviet orthodoxy. It is our intention in this book to provide a series of maps that will enable the reader to follow various phases in the development of Russian history from earliest times to the present. The accompanying text aims to provide a context for the actions and events shown on the maps, and is largely descriptive. The introductions preceding each section provide the larger picture, giving additional background information and occasional reference to historical controversies. Though not intended to be a comprehensive introduction to Russian history, those fascinated by recent events may find something in the past that helps them understand the complexities of the present.

Such a work as this owes much to previous scholarly endeavours and could not be written without reliance on the studies and research of others. In virtually every respect this work is also a collective effort and in that regard we are indebted to all the production team, though special thanks should go to the editors, Caroline Lucas and Stephen Haddelsey, whose encouragement and chivvying (of late) has ensured completion. I am also especially grateful to the following individuals for their encouragement, advice, comments and administrative assistance: Roger Bartlett, Leslie Collins, Lindsey Hughes, Lesley Pitman and Caroline Newlove. The final result, as always, is the responsibility of the authors. Closer to home, thanks must go to Emily and Alice who, in the early stages, readily made available crayons and felt tips and helped with colouring and to Susanne for assistance in countless ways but generally for enduring the hectic times of the past few months and providing a supportive environment.

John Channon
University of London, 1995

I should like to express my gratitude especially to the following colleagues and friends at the University of Derby: to Sue Wall, for her kind offer of administrative help, to Paul Bingham and Dominique Davison who provided faithful support and encouragement during a frenetic period of writing, most particularly to the following members of our Centre for Identity and Armed Conflict in Europe: Ian Barnes, Ian Whitehead and Beata Polonowska, and finally to Tony Carty who has been a constant source of inspiration.

Robert Hudson
University of Derby, 1995

Contents

Timeline: 6th Century to 1358

RUSSIAN STATE	TERRITORIAL EXPANSION AND INTERNATIONAL RELATIONS	ART AND ARCHITECTURE	RELIGION AND THOUGHT
6th–8th Century: migration of East Slavs from Central Europe to Russian steppe	860 first Slav attack on Constantinople		
c. 862 Viking traders (the Rus) form first Russian state in Novgorod with Rurik as head			c. 860 Cyrillic alphabet invented by Cyril and Methodius
c. 882 Oleg becomes first ruler at Kiev; Novgorod and Kiev united	911 first treaty with Constantinople		
913 Igor			
945 Svyatoslav I	944 second treaty with Constantinople		
973 Yaropolk I	967 Svyatoslav I sacks Khazaria		c. 955 Igor's widow Olga of Kiev is baptized
978 Vladimir the Saint			988 Vladimir begins mass conversion of his people to Orthodox Christianity
1015 Svyatopolk I			1015 martyrdom of Boris and Gleb
1019 Yaroslav the Wise	1025 Yaroslavl founded		
1054 Izaslav I		1037 Cathedral of St Sophia, Kiev	
1073 Svyatoslav II		1043 Cathedral of St Sophia, Novgorod	1051 Ilarion elected first native Russian metropolitan of Kiev and writes *On the Law and Grace*
1077 Izayaslav I			
1078 Vsevolod			
1093 Svyatopolk II			
1113 Vladimir Monomakh	1108 Vladimir founded	1113 Church of St Nicholas, Novgorod, early example of "onion" dome. *Primary Chronicle* completed	c. 1100 Vladimir Monomakh's *Testament*
1125 Mstislav I		1125 Virgin of Vladimir icon commissioned in Constantinople	
1132 Yaropolk II			
1146 Izayaslav II			
1149 Yurii Dolgorukii the Longsighted			
1154 Rostislav			
1157 Andrey Bogolyubsky	1156 Moscow founded		
1176 Vsevolod III of the Large Nest		1158 Dormition Cathedral, Vladimir	
1212 Yury II		*c.* 1187 *Saga of Igor's Campaign* 1198 Church of the Saviour, Novgorod	
	1221 Nizhny-Novgorod (Gorky) founded 1223 first Tatar raid 1237 end of Kievan Rus: Batu Khan's Tatar armies plunder cities and two centuries of vassalage to the Tatar Khanate of the Golden Horde begins		
1238 Yaroslav II	1242 Alexander Nevsky defeats Teutonic Knights on frozen Lake Peipus		
1247 Svyatoslav III			
1249 Andrei II			
1252 Alexander Nevsky			
1263 Yaroslav III ends Kievan line and Daniel begins Muscovite line	1270 Novgorod signs treaty with Hanseatic League		
1303 Yury III		1285 Transfiguration Cathedral, Tver	1300 Metropolitan see moved from Vladimir to Kiev
1305 Mikhail obtains *yarlik* (authorisation) from Golden Horde and is first to adopt title "Grand Prince of All Russia"			
1318 Mikhail executed and replaced by Yurii			c. 1321 Metropolitan see moved to Moscow
1325 Ivan Moneybags			c. 1340 Trinity monastery founded by St Sergius of Radonezh
1341 Semyon the Proud			
1353 Ivan the Fair			

Timeline: 1359 to 1600

RUSSIAN STATE	TERRITORIAL EXPANSION AND INTERNATIONAL RELATIONS	ART AND ARCHITECTURE	RELIGION AND THOUGHT
1359 Dmitry Donskoi		1378 Greek artist Theopanes comes to Novgorod and both influences and is influenced by Russian art	
1389 Vasily I	1380 Dmitry defeats Tatars at Kulikovo		
	1392 Annexation of Nizhny Novgorod and Suzdal by Moscow		
		c. 1400 *Trinity Chronicle* 1405 Iconostasis of Annunciation Cathedral, Moscow 1411 Rublyov's Trinity icon	
1425 Vasily the Blind 1425–50 Civil War; throne claimed by Yury of Zvenigorod			1429 Solovki monastery founded
1462 Ivan the Great	1462 end of Tatar Yoke. Ivan the Great begins expansion of Muscovite Russia by annexing neighbouring princedoms		1454 rival Orthodox metropolitan formed by Lithuanians in Kiev
	1478 Ivan crushes Novgorod 1480 Ivan ends tribute to the Tatars	1475–79 Dormition Cathedral, Moscow, built by Fioravanti	
	1494 Ivan closes Hanseatic depot in Novgorod	1491 "Gate of Salvation", main entrance to Kremlin, built	
	1500 Ivan attacks Lithuania		
1505 Vasily III	1503 truce between Muscovy and Lithuania		1503 Church council backs "Josephites" (followers of Joseph of Volokolamsk)
	1507 Vasily III attacks Lithuania		
	1510 Vasily III annexes Pskov		
	1514 Muscovy captures Smolensk		1516 cosmopolitan scholar Maxim the Greek comes to Moscow and is jailed for 20 years for attempting to introduce Italian Renaissance culture to Russia
	1522 truce between Muscovy and Llthuania	1532 Church of the Ascension, Kolomenskov	
1533 boyar rule during minority of Ivan IV 1547 Ivan the Terrible is first "tsar"	1547 Ivan the Terrible begins annexation of Tatar kingdoms to the east: this opens the door to Siberia and to Russia's future as a multiracial empire		
1549 first national assembly			
1550 new law code, the *Sudebuik*	1552 Ivan the Terrible conquers Kazan Khanate		1551 Hundred Chapters Council (Stoglav)
	1555 Muscovy Company formed in London	1555–60 Basilica of St Basil built in Red Square to celebrate Ivan the Terrible's capture of Kazan	
	1558 Ivan the Terrible occupies Estonia: Livonian War begins		
	1564 Peace between Muscovy and Sweden. Prince Kurbskii of Muscovy defects to Lithuania: begins repression in Muscovy		1564 *The Apostle* is first book to be printed in Moscow
1565 *Oprichnina* (army of licensed gangsters) established by Ivan the Terrible			
1567 failed conspiracy against Ivan			1569 Liublin Union of Poland and Lithuania strengthens Roman Catholicism and poses threat to Russian Church
1570 *Oprichniki* sack Novgorod			
1575 Ivan installs "parody tsar" and establishes second *Oprichnina*	1571 Tatars sack Moscow		
	1577 Livonia conquered by Ivan's Muscovite army		
	1579 Stefan Bátory of Poland declares war on Muscovy		
	1582 Muscovy occupies Siberia. Truce between Russia and Poland 1583 truce between Sweden and Muscovy		
1584 Fyodor I			1589 Moscow Patriarchate established
			1596 Union of Brent
		1596 Kremlin built at Smolensk	
1598 Boris Godunov is last Muscovite ruler; "Time of Troubles" (succession crisis) begins		1600 Church of St Nicholas, Panilovo	

Timeline: 1601 to 1712

RUSSIAN STATE	TERRITORIAL EXPANSION AND INTERNATIONAL RELATIONS	ART AND ARCHITECTURE	RELIGION AND THOUGHT
1605 Fyodor II becomes Tsar but in murdered; False Dmitry I pretends to throne 1606 Pretender killed in coup organised by Vasily IV, who becomes Tsar 1607 suppression of uprising by Ivan Bolotnikov			
1608 failed siege of Moscow by False Dmitry II 1610–13 Interregnum	1609 Alliance between Sweden and Muscovy 1610 Swedish–Muscovy forces defeated by Polish Army. Sigismund III of Poland abducts Vasily IV and offers throne to own son, Wladislaw 1612 Surrender of Polish garrison in Moscow		
1613 Mikhail is first Romanov ruler	1618 Truce between Poland and Muscovy	1618 "Marvellous Church" at Uglich	
	1625–35 Cossack revolts		
	1632 Muscovy attacks Poland in succession crisis following Sigismund III's death		
	1634 Peace treaty between Muscovy and Poland. Vladislav renounces claim to tsardom		1634 first Russian grammar book printed
			1640 Zealots of Piety, led by Tsar's confessors, formed to purge Church of impurities
1645 Alexis I			
1649 New legal code (*ulozheniye*) consolidates enserfment of peasantry	1648 Cossack revolt led by Bogdan Khmelnitsky	1649 Church of Nativity at Putinki, Moscow	
	1651 Khmelnitsky accepts rule by Ottoman sultan, but Polish forces defeat Cossacks and interim peace is imposed 1652 Cossacks refuse peace treaty with Poland and transfer allegiance to Muscovy		1652 Nikon becomes Patriarch and receives full authority to reform Church
	1656 truce between Muscovy and Poland. Muscovy attacks Swedish Livonia 1657 death of Khmelnitsky 1658 Cossack splinter group signs treaty with Poland 1659 rival Cossack group signs treaty with Alexis I. Truce between Muscovy and Sweden 1660 Polish victory over Muscovy at Cudnów	1667–70 Palace of Kolomenskov	1658 Alexis I breaks with Nikon, who resigns 1666 Nikon exiled by Church Council but his reforms are implemented. Avvakum leads Old Believers against reform and subjection of Church to state
1676 Fyodor III			*c.* 1670 *The Life of Archpriest Avvakum by Himself*
1682 Peter I shares tsardom with feeble stepbrother Ivan I but Sophia is effective ruler	1686 treaty with Poland confirms Russian possession of Kiev		1681 Avvakum burnt at the stake
1689 Peter the Great becomes sole tsar and reveals interest in sailing and shipbuilding	1689 Treaty of Nerchinsk with China		1687 Slavono-Graeco-Latin Academy formed in Moscow
1697–98 Peter begins Westernisation of Russia		1693 Church of Intercession of Holy Virgin at Fili near Moscow	
	1700 Great Northern War against Sweden begins		1701 School of Mathematics and Navigation founded in Moscow
	1703 site of St Peterburg captured from Swedes: allows empire access to Baltic Sea		1703 first Russian newspaper published in St Petersburg
1712 Peter transfers seat of government to St Petersburg	1712 Peter the Great defeats Charles XII of Sweden at Poltava		

Timeline: 1713 to 1862

RUSSIAN STATE	TERRITORIAL EXPANSION AND INTERNATIONAL RELATIONS	ART AND ARCHITECTURE	RELIGION AND THOUGHT
		1714 Church of Transfiguration, Kizhi	
1721 Peter becomes first emperor	1721 Peace of Nystadt ends Northern War	1722 Palace of Petrodvorets, St. Petersburg	1721 abolition of Patriarchate
1725 Catherine I (Peter's widow)			1725 Academy of Sciences set up in St Petersburg
1727 Peter II			
1730 coup by Peter's cousin Anna after his death			1731 Noble Military Academy set up in St Petersburg
1741 Elizabeth banishes regent Anna, deposes infant Ivan IV and claims throne		1754 Rastrelli's Winter Palace, St Petersburg	1752 Lomonsov's *Letter on the Utility of Glass*
			1755 Moscow University opened
			1757 Academy of Arts founded, St Petersburg
1762 Peter III ascends throne and liberates gentry from obligatory service. His wife Catherine has him put to death and claims the throne	1762 Catherine the Great begins extension of Russian sovereignty to west to absorb most of Poland and Lithuania. Vast numbers of Russian peasants and German immigrants are settled in Ukraine and along the Volga	1766 Falconet begins Bronze Horseman, St Petersburg	1763 Catherine the Great decrees the introduction of a modern system of general schools
1767 Legislative Commission			
1773 rebel Cossack chieftain Pugachev proclaims himself Peter III and sets up bogus court		1778 First performance of Bortniansky's *Creon*	
1774 Pugachev beheaded and quartered			
1785 Charter of Nobility by Catherine the Great	1784 annexation of Crimea		1786 Statute of Popular Schools proposes institution of primary and secondary schools
			1790 Radishchev's *Journey from St Petersburg to Moscow*
1796 Paul I	1799 Sitka founded as capital of Alaska		
1801 Paul murdered in palace revolution and succeeded by son Alexander I	1807 Treaty of Tilsit signed with Napoleon	1806 Old Admiralty building remodelled	1802 Ministry of Education set up
	1812 Napoleon invades Russia, aiming to spark peasant uprising but French forced to retreat to Paris. More than 90% perish en route	1812 *Trinity Chronicle* destroyed in fire but reconstructed by scholars	1816–26 Karamzin's *History of the Russian State*
1825 Decembrist uprising against Romanov autocracy by young army officers. Nicholas I becomes Tsar	1822–55 Nicholas I helps Greece in her war of independence	1831 Golden Age of Literature begins: Pushkin's *Eugene Onegin* is first novel to take contemporary society as subject of fiction	1830–03 Speransky's 51-volume *Complete Collection of Laws of the Russian Empire*
	1832 Duchy of Warsaw becomes part of Russia	1836 Glinka's *A Life for the Tsar* founds a national school of opera	1835 University Charter
			1836 Chaadayev's first *Philosophical Letter*
		1840 Lermontov's poem *Hero of our Time*	
		1842 Gogol carries the new realism further with *Dead Souls*	1848 15 members of Petrashevtsy intellectual group which included Dostoyevsky sentenced to death for "conspiracy of ideas"
1855 Alexander II	1853 outbreak of Crimean War	1849 Moscow Kremlin completed	
	1860 Vladivostok founded	1859 Goncharov publishes *Oblomov*	1860s ratio of literates to illiterates is 1:15. Bookshops multiply sixfold and many libraries opened
1861 pressure from all levels of society leads to emancipation of serfs		1862 Turgenev publishes *Fathers and Sons*	1861 Disturbances at universities following emancipation

Timeline: 1863 to 1943

RUSSIAN STATE	TERRITORIAL EXPANSION AND INTERNATIONAL RELATIONS	ART AND ARCHITECTURE	RELIGION AND THOUGHT
		1864–69 Tolstoy's *War and Peace*	
		1866 Dostoyevsky's *Crime and Punishment*	
	1867 Alaska sold to U.S.A.		1872 Bakunin expelled from First International but has influence on later revolutions
		1873 Rimsky-Korsakov's *Ivan the Terrible* first performed	
		1874 Mussorgsky's *Boris Godunov*	
			1879 Land and Freedom revolutionary group splits into People's Will and Black Partition
1881 Alexander assassinated by People's Will and succeeded by son Alexander III, who wreaks revenge by persecuting Jews	1891 building of trans-Siberian railway begins	1890 Borodin's *Prince Igor*	
		1892 Rachmaninov's *Piano Concerto No. 1*	
1894 Nicholas II			
		1897 Moscow Arts Theatre founded	1897 ratio of literates to illiterates is 1:5
		1898 first performance of Chekhov's *The Seagull*	
		1902 Gorky's *The Lower Depths*	
1905 revolution in St Petersburg and Moscow put down by force but leads Nicholas to set up a parliament (Duma) and grant civil liberties	1904–05 Russia defeated in Russo–Japanese War	1906–13 Gorky exiled	1912 first issue of *Pravda*
	1914 Russia enters World War I against Germany. In the Battle of Tannenburg Russian troops invading East Prussia are defeated by Germans	1914 Stravinsky's *The Rite of Spring*	
1916 faith-healer Rasputin murdered by aristocrats to end his influence over the Tsarina		1916 Bunin's *Gentleman from San Francisco*	
1917 February and October revolutions force Tsar's abdication. Lenin seizes power from provisional government		1917 Kandinsky paints *Ladies in Crinoline*	
1918 Nicholas II and family murdered	1918 Treaty of Brest-Litovsk ends Russian participation in World War I. The Bolshevik régime renounces all claims to Poland, Ukraine, Lithuania, Finland and the Baltic provinces	1918 Malevich's Suprematist canvas *White on White*	1918 Patriarchate re-established
1918–20 Civil War between Bolsheviks and White Russians	1919 Third International founded in Moscow with aim of stimulating worldwide revolution	1919 Mayakovsky publishes *Mystery Buff*	
1921 New Economic Policy			1922 confiscation of Church property
1922 Stalin becomes secretary-general of Communist Party central committee			
1923 USSR formed; Russia is largest republic, with Lenin as leader			
1924 Death of Lenin			
		1925 Eisenstein's film *Battleship Potemkin*	1925 death of Tikhon: no new patriarch elected
	1927 20 alleged British spies executed in Moscow	1926 Shostakovich's *Symphony No. 1* and Sholokhov's *And Quiet Flows the Don*	
1928 end of NEP and introduction of Five Year Plans; forced industrialization and collectivization begins			1930s Cultural Revolution
		1932 artistic groups officially dissolved	1932 primary education made compulsory
		1933 Bunin receives Nobel Prize for Literature	
	1934 USSR joins League of Nations	1934 First Congress of Union of Soviet Writers: Doctrine of Socialist Realism formulated	
1936–38 Great Purges: Stalin kills millions of his opponents and sends millions to gulags		1936 Prokofiev's *Peter and the Wolf*	
	1939 Non-Aggression Pact with Germany		
1940 Trotsky murdered. Stalin is official head of government		1940 Anna Akhmatova begins *Poem Without a Hero*	
	1941 Hitler invades USSR		
			1943 Church leaders meet Stalin: Patriarchate re-established

Timeline: 1944 to 1993

RUSSIAN STATE	TERRITORIAL EXPANSION AND INTERNATIONAL RELATIONS	ART AND ARCHITECTURE	RELIGION AND THOUGHT
	1945 World War II ends: 25 million Russian lives lost, and 100 million East Europeans drawn into Soviet bloc 1947 era of Cold War begins	1946 Akhmatova and others expelled from Union of Soviet Writers	
			1952 Stalin's *Economic Problems of Socialism in the USSR*
1953 death of Stalin, Malenkov is unofficial successor 1955 Khrushchev is first secretary of central committee and Bulganin is head of state		1954 Erenburg's *The Thaw* is first book to refer to horror of Stalin's regime. Receives reprimand from Writers' Union	1954 government steps up campaign against Church
1957 launch of Sputnik I	1956 Hungarian uprising against Soviet occupation is crushed	1957 Pasternak's *Doctor Zhivago* 1958 Pasternak declines Nobel Prize for Literature	
1961 Gagarin is first man in space	1962 Cuban missile crisis brings threat of nuclear war. Khrushchev withdraws missiles on Kennedy's insistence	1962 Solzhenitsyn's *One Day in the Life of Ivan Denisovich*	
1964 Brezhnev ousts Khrushchev in coup		1965 Sholokhov receives Nobel Prize for Literature	
	1968 "Prague Spring": Czechs resist invasion by Red Army		1966 writers Sinyavsky and Daniel sent to labour camps for disseminating anti-Soviet propaganda. Law passed restricting Church activity
		1970 Solzhenitsyn receives Nobel Prize for Literature	1972 Brodsky exiled 1974 Solzhenitsyn exiled
	1979 Soviet invasion of Afghanistan	1980 Brodsky's *A Part of Speech*	
1982 Andropov succeeds Brezhnev 1984 Chernenko succeeds Andropov 1985 Gorbachev succeeds Chernenko and expounds *glasnost* (openness) and *perestroika* (restructuring) 1986 nuclear disaster at Chernobyl	1986 Gorbachev and U.S. president Reagan sign first treaty on inspection and reduction of nuclear weapons	1987 Brodsky receives Nobel Prize for Literature	
1988 Gorbachev proposes institution of elected legislatures and a presidency	1988 USSR agrees to pull troops out of Afghanistan 1988 New trans-Siberian railway, the Baikal–Amur Mainline, built		
1990 Gorbachev becomes president in first freely-elected parliament (Supreme Soviet) and wins Nobel Peace Prize. Soviet economy collapses and Gorbachev granted special powers 1991 Yeltsin defeats coup against Gorbachev by Communist hardliners and suspends Communist Party. Formal dissolution of USSR as Latvia, Lithuania and Estonia proclaim independence. Yeltsin remains president in first free elections but there is a shift to the right when Zhirinovsky is voted overall leader 1992 Yeltsin dissolves parliament and announces he is taking control. Political violence leads to state of emergency. Communist hardliners stage a violent but unsuccessful siege 1993 Yeltsin dissolves Supreme Soviet but shift to extreme right continues in elections and key pro-western reformers resign. Nationalist violence intensifies.	1990. Reduced global tensions are replaced with nationalistic tensions; Gorbachev sends troops into Vilnius when Lithuania secedes from USSR	1991 Solzhenitsyn's treason charge dropped: he later returns to Russia. Brodsky made US poet laureate	1993 Zhirinovsky's *The Last Thrust to the South*

I: The Origins of Russia

From obscure beginnings, the various peoples of Russia gradually extended their territory and through military conquest and trading ventures sought an increase of prestige and power.

"Oleg settled as prince in Kiev. And he said. 'Let this be the mother of the towns of Rus.' He had Varangians and Slavs and all the others with him. And from that time on they were called Rus."
Primary Chronicle

In the 7th and 8th centuries the Slavonic tribes chiefly inhabited territories now known as Belarus and Ukraine. Their neighbours to the north were the Finnic-speaking Baltic peoples; to the west were Slavonic tribes later known as Poles; while the lands to the northeast—territories which the Russians conquered and settled later in the Middle Ages and which subsequently became known as the heart of Russia—were then populated mainly by Finnish tribes. To the east and south, in the steppe lands between the River Dnieper and the River Volga, lived nomadic peoples, mostly Turkic-speaking (Bulgars, Khazars), and the Magyars. The latter were later succeeded by the Turkic-speaking Pechenegs and Polovtsians whose terrible raids were to threaten the survival of the medieval Russian state. At this time the nearest great powers were the Khazars, an Asiatic-Turkic people who settled in the Lower Volga, the North Caucasus and the southeastern steppe, and the Bulgarian empire in the Balkans—created in the 680s by some of the Bulgar nomads from the "south Russian" steppe. The greatest power of the time, which had possessions in the Crimea, was the Byzantine Empire which ruled part of the Balkans and Asia Minor. The capital of the Byzantine Empire was Constantinople.

Very little is known about the previous history of the areas inhabited by these Slavonic tribes. It is understood that the "Cimmerians" ruled an area to the north of the Black Sea from about 1000 to 700 BC. They were succeeded first by the Scythians, and then, around 200 BC, by the Sarmatians—Iranian nomads who spoke an Indo-European language—who continued the east–west trade route through the southern steppe. These peoples are mentioned in the works of the first known historian, the Greek Herodotus.

From AD 200 a number of tribes and peoples invaded the northern Caucasus. Between AD 200–370 the north Black Sea coastal region was settled by German-speaking people called Goths. The Goths were driven into central Europe by the Huns, who swept across the Volga and eventually crossed into central and western Europe. It was during the short-lived empire of the Huns that Turkic-speaking nomads were introduced to eastern Europe. Following the collapse of this empire in the 450s, the steppe lands of eastern Europe were dominated by the nomadic powers of the Bulgars and Khazars. For a brief period in the mid-6th century "southern Russia" was controlled by Tatars, and probably Turkic-speaking nomads called Avars, who soon moved westward to establish a great empire in what is now Hungary.

After a brief period of Bulgar power, the steppe peoples and many of the Slavonic tribes inhabiting "Ukraine", passed under the control of the Khazars. Using the great rivers that traversed these areas, the Khazars controlled the trade routes between the Volga and the Don, the isthmus between the Caspian and Black Seas and the Kuban steppe, subjugating the Slavs who had settled around the Dnieper in the 7th century. Khazar prosperity was a result both of tribute and their strategic position astride the trade routes. The Khazars are usually credited with bringing stability and religious freedom to the region—and in the 8th and 9th centuries the élite converted to Judaism. The conquests of the Khazars caused the division of the Turkic-speaking Bulgars into two groups: one group established its capital at Great Bulgar, on the confluence of the Volga and Kama rivers, and subsequently converted to Islam; another group established itself in the Balkans and created the medieval state of Bulgaria. Khazar power was broken eventually by one of the rulers of the first Russian state, Svyatoslav I (*c.* 950s), although the origins of Svyatoslav's ancestors and the way in which the Slavonic tribes passed under their control remains a subject of some controversy.

In the 7th and 8th centuries the Volga, a major trade artery to the East, began to attract Vikings (Varyagi or Varangians in Russian) from the north and northwest who settled and controlled the Baltic. While they had explored the region in the 6th and 7th centuries, the next two centuries were to witness more aggressive Varangian incursions, their aim being to develop a trade route (for honey, wax, furs, amber and slaves) from the Gulf of Finland to Kiev and then, via the Dnieper and the Black Sea, to Constantinople. In the 9th century the grouping of disparate cultural and political forces—comprising the east Slav tribes—joined to form what became known as Kievan Rus. The *Primary Chronicle*, written some 300 years after the event, explains that the Varangians exacted tribute from the Khazars but in 862 the Varangians were forced out and Slavs began to rule themselves. The so-called "Normanist" theory maintains that the origin of Rus is Scandinavian; Scandinavian names, such as Oleg, Olga, Igor are mentioned in the Treaty of Kiev between Rus and Byzantium in 911. Similarly, place names in north central Russia, including *Moskva* (Moscow), derive from Finno-Ugric. Soviet historians, however, argue that Rus referred to Slavs and not Varangians and that there was no Scandinavian influence on early Russian language, literature, religion or law. Furthermore, the Greek "Rhos" was familiar in southern Rus prior to the Varangians, "Rus" in fact referring to the area around Kiev—where there is also a river named "Ros"—and not the area around Novgorod where they first appeared. While Scandinavians plundered and traded, they were also absorbed into the population, making it difficult to find traces of them later. It seems probable that Varangians established Kievan Rus in the mid-9th century, that trade with Constantinople was their main object and that they needed Kiev as a staging post.

Between 880 and 890 the name Rus was applied to a new loose political unitary federation, the largest of its kind in medieval Europe. The military power of Kievan Rus was greatly feared by its neighbours. The campaigns of Svyatoslav, who ruled 960–970, took him east, west and south, his aim being to control the trade routes. Kiev was ideally situated for trade, exporting slaves to Prague, one of the centres of southern trade at this time. Soon Kiev conquered the entire Volga River network, ruling from the Baltic to the Adriatic. Svyatoslav defeated the Bulgars and Khazars, then focussed his attention on Bulgaria, which had accepted Christianity in 864. Byzantium had paid tribute to Bulgaria since the 680s but in the 960s, with its power in decline and wracked by internal dissent, Byzantium decided to attack, and encouraged Svyatoslav to do likewise. Svyatoslav attacked via the Crimea and Black Sea and succeeded in conquering Bulgaria, then remained there as the new ruler. Realising that Svyatoslav planned to settle in Bulgaria permanently, the Byzantine emperor incited the Pechenegs (Asiatic nomads) to attack Kiev. Although Svyatoslav thwarted the attack, he returned to Bulgaria in 969 and proclaimed himself conqueror. In 971 Svyatoslav invaded Byzantium but was defeated; on the return journey from Byzantium to Kiev, he was killed by the Pechenegs.

The funerary statue of a Pecheneg leader. It was the Pecheneg commander Kurya who ordered that the conquered Svyatoslav be beheaded and his skull be lined with silver for use as a ceremonial drinking vessel.

From the mid-10th century Kiev was a federation of principalities, unified through the centre. Local princes ruled their own states with allegiance to the Great Prince of Kiev. However, this did not prevent local princes challenging the Great Prince; by the 12th century the system had begun to fall apart, culminating in 1132–35 with the fragmentation of the federation into more than 12 principalities, with Kiev as the capital.

From the perspective of the unitary and centralized Muscovite state, Kievan Rus appeared to offer greater freedom. It was a federation of city states each with its own prince and aristocracy. While not an autocrat, the prince was responsible for military defence, justice and protection of the Church. Before decisions were made, the prince had to consult with the boyars (the aristocracy). An inner Council met every day. Boyars were not vassals of the prince and could serve another prince if they wished. By the second half of the 11th century the growing wealth and influence of cities and the influence of ordinary citizens, increased. The *veche* (popular assembly) acting as a powerful check on the prince, had now become an important institution. While slaves (usually war captives) were bought and sold, half the citizenry were legally free and could inherit land, while the mixed composition of urban society included free merchants.

In 990 Rus converted to Christianity. Olga, mother of Svyatoslav I, who had acted as regent when her husband, Prince Igor (913–945) was killed, had been baptised a Christian, but had failed to convert Svyatoslav. Svyatoslav's son, Vladimir, converted to Christianity in 988–89. Historically, Christianity became significant because of the west European tradition of marriage alliances. Essentially the Church—acting as a unifying factor—held the country together. It also exerted a powerful cultural influence. Accepting the eastern, or Byzantine, form of Christianity clearly affected Russia's subsequent development.

In the early 12th century the power of Kiev began to wane. After the death of Vladimir II, Monomakh (reigned 1113–1125), the Kievan federation finally disintegrated. The collapse was due to an increase in the power of other principalities, separatist tendencies, rivalry between princes and the role of the *veche* and the boyars. The Polovtsy (or Cumans) cut the Kievan trade route to the south, though the Byzantine–Kiev trade route was already declining in importance. Kiev lost its pre-eminence, becoming just another regional trading centre. By mid–late 12th century, Kievan rule was over and a new power struggle had developed.

In 1169 Prince Andrei Bogolyubsky of Suzdal sacked Kiev, becoming the most powerful prince. Making Vladimir his base, which became the centre of Rus, he then returned north where he founded his own village of Bogolyubovo. Supremacy in northern Rus meant control of Novgorod, a unique merchant city with long established trade links in the Baltic. The city was governed in the 12th century by the *veche*. There was no ruling dynasty of princes; indeed, princes were virtually hired and had to sign a contract. High positions, such as that of the bishop, were elected by the *veche*, and acknowledged the overlordship of Vladimir.

Vladimir was a worthy successor to Kiev; it possessed a cathedral (a splendid example of early Russian architecture), its own school of icon painting and was an important trading centre. Despite these indications of prestige, in comparison with his Kievan predecessors, the Prince of Vladimir had less authority over the other, minor princes. This diminution of power was largely due to decentralization and the proliferation of small principalities under the appanage system. This system involved the grant of lands to junior branches of the family in return for obedience. After death, the land would return to the prince who maintained the right to dispose of it as he wished.

In the 13th century Russia was polarized into the northeast and southwest. Enjoying closer links with the West and acting as middle-men in the trade between the West and the Baltic, Volhynia and Galicia became dominant in the southwest division. As a border territory, this region became the scene of wars and devastation but did enjoy unity for a short period.

The ill-fated princes Boris and Gleb who were murdered on the orders of their brother Svyatopolk. Russia's first Christian martyrs, Boris and Gleb were venerated for their humility in facing the ruthless ambition of their sibling.

From the late 13th century Prince Daniel was in control of Volhynia and reunited southwest Rus. Despite having fled before the approaching Tatars in 1240, Daniel returned immediately after their departure. In order to stem the advance of the Teutonic Knights, in 1245 Daniel accepted the crown from the Prince of Rome. At the end of the 13th century the Tatars again appeared. Daniel died in 1264 and the Tatars continued to follow their traditional policy of "divide and rule". Galicia and Volhynia were swallowed up by Poland and Lithuania, and power in Russia again passed to the northeast until the 17th century.

Origins of the Slavs

Pushing eastwards into what is now central Russia, the Slavs settled in separate groups along the Dnieper, an area known to them as Rus.

> *"Slavs scattered about the land and took names for themselves, depending on the place in which they settled … And in this way the Slavonic nation spread."*
> Primary Chronicle

Russians are descended from one of three branches of the eastern Slavs—populating areas now known as Russia, Ukraine and Belarus—all of whom were united by a common Slav language. By 800 BC these east Slavs, who came possibly from the Caucasus though their origin is uncertain, were spreading eastwards from Europe into the woodlands of what is now central Russia. They appear to have settled in a number of separate groups along the River Dnieper and along the Vistula and the Don, spreading over an area they knew as Rus.

After 800 BC, under increasing pressure from China, a succession of nomadic tribes and horsemen moved from Asia westwards across the southern steppe. One of these tribes, the Scythians—which settled north of the Black Sea—by 600 BC had probably subjected some of the Slavs to their rule, and driven others further and further northwards. The Scythians were divided into many small groups, or "hordes", each recognizing the authority of the king of the main horde. While their use of the saddle gave them military supremacy over the Greeks, who rode bareback, both peoples proved willing to trade peacefully with each other—Scythian cattle, hides, furs, timber, wax, honey and grain being exchanged for Greek olive oil, wine and textiles.

Around 200 BC, another Asiatic tribe, the Sarmatians, extended into what is now southern Russia, where they held sway for over 400 years. During this time, the Roman Empire absorbed the Greek colonies (founded *c.* 700 BC) on the Black Sea and spread to the western shore of the Caspian Sea.

A gold plaque manufactured by the Sarmatians portrays horsemen and their mounts at rest in the shade of a tree. Both the Scythians and the Sarmatians were of Iranian origin.

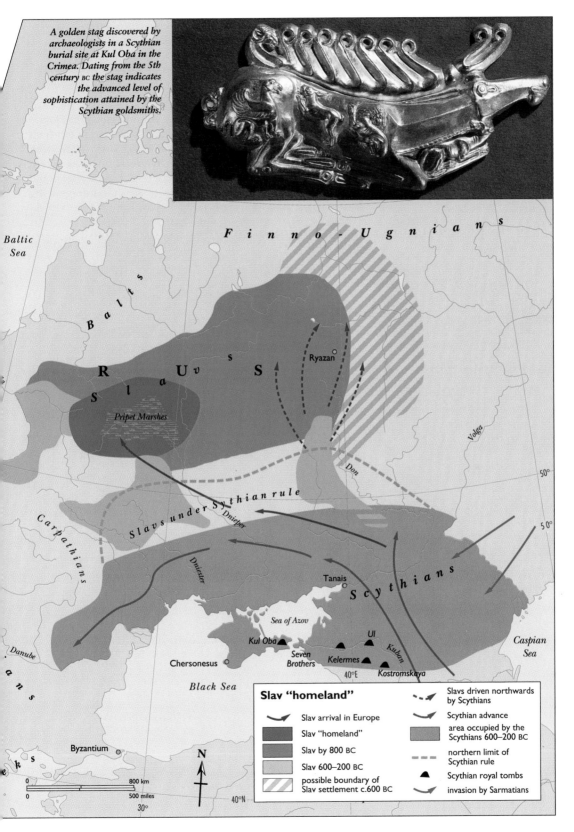

A golden stag discovered by archaeologists in a Scythian burial site at Kul Oba in the Crimea. Dating from the 5th century BC the stag indicates the advanced level of sophistication attained by the Scythian goldsmiths.

Baltic Sea

Finno - Ugnians

Balts

R S l a U v s S

Pripet Marshes

Ryazan

Volga

Don

50°

Carpathians

Slavs under Scythian rule

Dnieper

50°

Dniester

Tanais

Scythians

Sea of Azov

Kul Oba

UI

Kuban

Seven Brothers

Kelermes

Caspian Sea

Chersonesus

40°E

Kostromskaya

Danube

Black Sea

Byzantium

Slav "homeland"

⌒ Slav arrival in Europe	⇢ Slavs driven northwards by Scythians
▓ Slav "homeland"	⌐ Scythian advance
▓ Slav by 800 BC	▓ area occupied by the Scythians 600–200 BC
▓ Slav 600–200 BC	--- northern limit of Scythian rule
▨ possible boundary of Slav settlement c.600 BC	▲ Scythian royal tombs
	⌐ invasion by Sarmatians

N

0   800 km
0   500 miles

30°

40°N

19

Expansion of the Slavs

Between AD 200 and 750 the Slavs experienced many changes in their fortunes, sometimes subjugated by powerful neighbours, at others harassing them in return.

"Clan rose up against clan, and there was strife between them. Our land is great and rich, yet there is no order in it."

Primary Chronicle

In AD 200, the Goths defeated the Sarmatians and extended their rule to the Black Sea. Half a century later they defeated the Roman Emperor Decius on the lower Danube. After 300, the Goths were converted to Christianity by which time their overlordship was recognized by the majority of Slav tribes. By 360, the Goths had been driven westwards by the Huns, and by 451 the latter dominated most Slav tribes from the Volga to the Baltic and the Rhine. During the next hundred years, the period of the first Slav expansion, the strength of the Huns waned rapidly. By 500 they had retreated to the Lower Don and Volga, while another group settled in the Crimea.

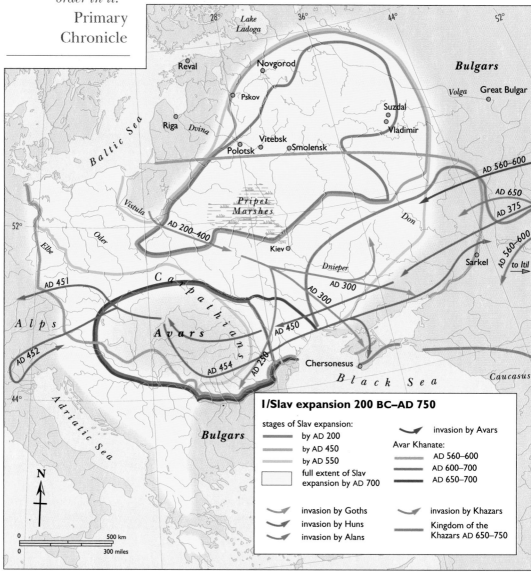

I/Slav expansion 200 BC–AD 750

stages of Slav expansion:
- by AD 200
- by AD 450
- by AD 550
- full extent of Slav expansion by AD 700
- invasion by Goths
- invasion by Huns
- invasion by Alans

- invasion by Avars

Avar Khanate:
- AD 560–600
- AD 600–700
- AD 650–700

- invasion by Khazars
- Kingdom of the Khazars AD 650–750

Once the Germanic tribes had crossed into Britain and France, the Slavs spread rapidly to the Elbe and Danube. The Huns of the Don formed the Khanate of Great Bulgaria, while the eastern Roman Empire retained precarious control along the shores of the Crimea, in the Caucasus and around the Adriatic.

An eastern Tatar tribe, the Avars, reached Europe after 550 and were used by the Byzantine Emperor Justinian to subjugate the Slavs, who were making frequent raids into the Balkans. In 562, the Avars reached the Elbe where they maintained power for 40 years. The century after 605 witnessed a Slav recovery: Avar control on the Elbe was thrown off, while other Slavs migrated into the Balkans. By 650, the Avars were confined to the middle and lower Danube.

The Khazars, another Asiatic tribe, settled on the shores of the Caspian, the lower Volga and the northern Caucasus, establishing the Khazar Kingdom (650–750). They split the Bulgars, one group settling in the Balkans and forming present-day Bulgaria; the other retreating north-eastwards, establishing its capital at Great Bulgar. The Khazars played a different historical role from that of the Huns and Avars; their wars against the Arabs prevented the spread of Islam into Europe, and they were noted for their international commerce, religious tolerance and enlightened laws. Although a semi-nomadic people, they promoted urban development, building towns like Itil (their capital) and Samander. Their prosperity was founded on their strategic position astride important trade routes, located at the crossroads of two continents.

A 3 metre tall stone representation of the Slav god Svantovit. Little is known about the pre-Christian religions of northern Europe, and the few details we do possess are frequently warped by the bias of the Christian chronicles.

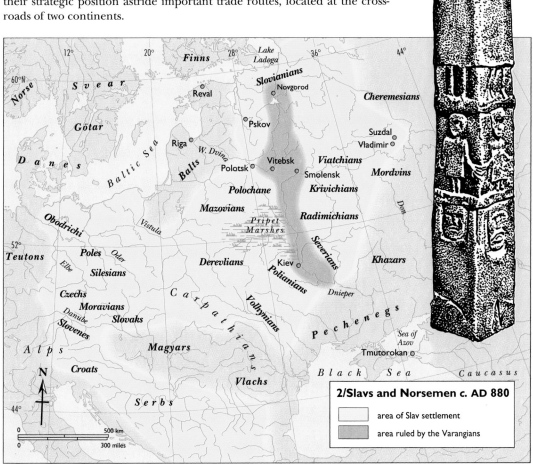

2/Slavs and Norsemen c. AD 880

☐ area of Slav settlement

▨ area ruled by the Varangians

Kievan Rus

With a culture based upon military prowess and trade, the Varangians became the dominant people of Russia.

"When a child is born to any man among them, he takes a drawn sword to the new-born child and places it between his hands and says to him: 'I shall bequeath to thee no wealth and thou wilt have naught except what thou dost gain for thyself by this sword of thine'."

Ibn Rusta,
an Arab scholar

In the 9th century, Varangian or Varyagi (the Slav term for Viking or Norse) merchant-warrior princes from the north settled in European Russia. It was they who developed Rus as a trading people, exploiting the great south-flowing rivers that link the Baltic with the Mediterranean, Black Sea and the Caspian. Originally based at Novgorod, the Varangians later moved their capital to the more strategically located city of Kiev, and established the loose federation known as Kievan Rus.

Russia's rivers, which ran through forest and steppe, were vital to the Varangians, but a political unit based on rivers was insufficiently strong to survive. Kievan Rus was dominated by constant struggle; it ultimately failed to retain its steppe territory which again fell under the control of westward migrating nomads.

The Varangians' first independent ruler, Oleg (reigned 882–912), concerned to protect the waterways against the threat from the east—specifically from the Khazars—initiated a policy of expansion and conquest. By about 912, the time of his death, Oleg's realm included the majority of eastern Slav tribes.

The reign of Oleg's successor, Igor (*c.* 912–945), witnessed a number of unsuccessful foreign campaigns and showed few permanent gains. The brief, active rule of his son, Svyatoslav, saw Kievan Rus's activity beyond its borders reach its zenith, creating what has been called "the first Russian Empire". While his victories were often spectacular, over-extending his forces brought Svyatoslav only fleeting triumphs. Although most of his new empire was lost after his death in 972, the territories retained by Rus constituted a considerable net gain over the area bequeathed to him by Oleg 60 years earlier.

A detail from a rare illuminated manuscript depicts the bearded Varangians in full armour, carrying unit standards, axes and spears.

In about 981, Svyatoslav's son, Vladimir ("the Saint"), seized Przemysl, Cherven and their surrounding areas from the Poles. Two years later he conquered the Yatvyagi, and in about 983 annexed the area between the middle Nieman and Western Bug rivers. Vladimir's son, Yaroslav, resumed the advance in the west at the expense of the Poles, Lithuanians and Finns. In 1030, he founded the town of Yurev (Dorpat) in the northwest, and subsequent gains in the region secured Novgorod's control of the southeastern littoral of the Gulf of Finland. At the time of Yaroslav's death, Kievan Rus was the largest federation in Europe, uniting all the eastern Slavs and including several non-Slavic tribes.

Kievan Rus, 880–1054

Rus c. 880–912

expansion of Rus, c. 912–972

expansion of Rus, c. 972–1054

area temporarily paying tribute to Rus

Khazar empire to 967

principal trade route

Svyatoslav's campaign

town sacked by Svyatoslav

battle

Principalities and Novgorod Republic

Ruled by Viking merchant-warriors, Kievan Rus was to expand by the 11th century to become the largest federation in Europe.

"The horses neigh beyond the Sula, the glory echoes at Kiev, the trumpets blare at Novgorod, the banners stand fast at Putivl." From the *Tale of the Armament of Igor,* late 12th century.

Shortly before his death, Yaroslav divided the realm of Kievan Rus among his five sons. This set a precedent that contributed to the intense dynastic feuding of succeeding generations, and eventually resulted in the rise of independent and often warring principalities. United briefly from 1113–1125 by Vladimir II (Monomakh), the Russian lands were again divided and in conflict during the 100 years before the Tatar invasion. Polovtsy (or Cuman) raids emptied the southern lands, but colonization of the forest increased the populations of the northern and central principalities, enabling them to throw off Kievan overlordship. Novgorod, which had built up a vast fur-trading empire, and Vladimir-Suzdal which contained the fast-growing commercial centre of Moscow, were the leading forest principalities. By the time of the Tatar invasion in 1237, Vladimir-Suzdal was out to challenge the Volga Bulgars whose stranglehold on the middle Volga obstructed further Rus expansion eastwards. Nizhniy-Novgorod was built as a first move in this campaign.

Border raids and the sacking of Constantinople by the Crusaders in 1204 were just two factors that contributed to the decline of Kievan Rus. The Polovtsy, who by the 12th century had succeeded the Pechenegs as the major threat in the south, devastated the steppes, forcing the disunited Rus to withdraw from the mouths of the Dnieper and Southern Bug, further hampering Kiev's commerce.

In the northwest, powerful new enemies appeared in the 13th century: the semi-religious German crusaders known as the Teutonic Knights, seized Yurev (Dorpat), threatened Pskov and Polotsk, and spread eastwards along the right bank of the Western Dvina.

While the realm as a whole lost cohesion during this period, sizeable gains were made by Vladimir-Suzdal and "Lord Novgorod the Great". By 1136, the latter had obtained complete independence from Kievan Rus. For over 300 years Novgorod was a flourishing trading and cultural centre, repelling attacks from Teutonic Knights, Swedes, Lithuanians and Tatars. In 1478 it

Vladimir Monomakh, Grand Duke of Kiev (1113–1125), in council with his advisors. Temporarily halting the movement toward disintegration, Monomakh combined the Russian principalities in common defence against the Polovtsian threat. This representation is taken from a wooden panel carved in 1551 on the imperial seat of Ivan the Terrible in the Church of the Assumption in Moscow.

was finally crushed by Ivan the Great, and annexed to Moscow.

By 1237 the Russian landscape had gradually been altered by the founding of almost 300 towns and cities, and the urban concentrations in the northeast and southwest are indicative of the population shifts away from the turbulent southeastern frontier.

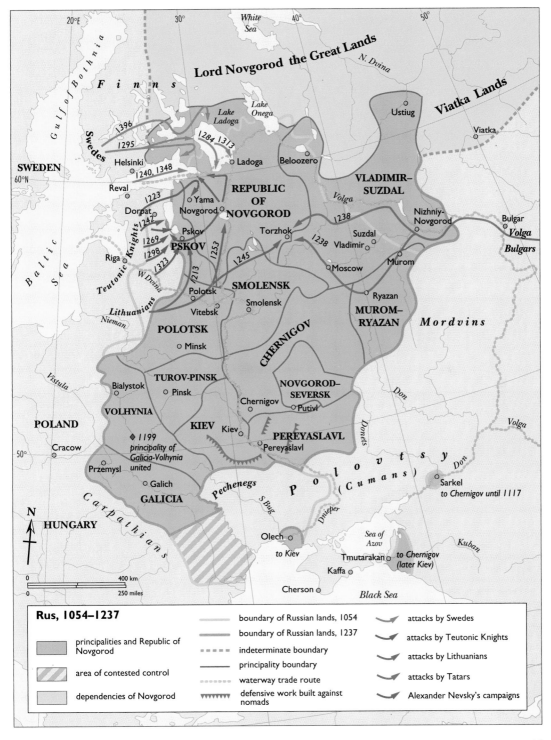

Rus, 1054–1237

- principalities and Republic of Novgorod
- area of contested control
- dependencies of Novgorod

- boundary of Russian lands, 1054
- boundary of Russian lands, 1237
- indeterminate boundary
- principality boundary
- waterway trade route
- defensive work built against nomads

- attacks by Swedes
- attacks by Teutonic Knights
- attacks by Lithuanians
- attacks by Tatars
- Alexander Nevsky's campaigns

Christianity in Russia

Founded in 988–89, the Russian Orthodox Church over the next 1000 years was to experience peaks of power and prestige but also troughs of savage repression.

"Vladimir, rejoicing that he and his subjects now knew God, looked up to heaven and said, 'O God ... look down on thy new people, and grant them, O Lord, to know thee as the true God.'"
Primary Chronicle

Shortly before AD 1000, following Vladimir's marriage to a sister of the Byzantine Emperor, Kievan Rus adopted Christianity. The spread of Christianity led to the division of the Slav world: while the Croats (in AD 700) and the Poles (in 999) were converted to Western (Roman) Catholicism, the Serbs (in 700), Bulgars (865) and Russians (988) were converted to Eastern (Orthodox) Catholicism under the authority of Constantinople. This led to strong antipathy between Russians and Poles, and also between Serbs and Croats.

The conversion to Christianity was particularly significant for the development of the arts in Russia. Metalwork, architecture and painting all flourished under Church patronage with much of the finest work bring inspired

Above: Up to the end of the 17th century wood remained the most dominant building material in Russian architecture. Although in the 10th century Novgorod boasted a 13-domed wooden cathedral, the vulnerability of these early wooden structures to decay and fire has resulted in the loss of all examples dating from before the late-medieval period.

Right: The Divine Trinity by Andrey Rublyov (1422–27). Usually painted in tempera on wooden panels, icons became an essential element of Russian Orthodox worship.

by the artists and craftsmen of Constantinople.

Russian monasticism flourished between 1200 and 1600, the years between 1200 and 1350 witnessing an intense development of urban monasteries. By 1400 the majority of monasteries to be founded was rural or "desert". Between 1350 and 1450, over 150 new monasteries were established, and by 1500, many monastic colonies had been set up in the predominantly pagan areas between Galich and the Urals. In 1588 the English Ambassador to Moscow noted that the monasteries owned all the best land in Russia.

Having thrived for over 900 years, Russian Orthodoxy—in common with all other religions within the USSR—was brought to the very brink of extinction under the repressive and atheistic Soviet regime. During the 1920s and 1930s, the Church and its adherents were actively persecuted, with innumerable churches converted to secular uses or completely destroyed. In Moscow repression was carried out with particular severity, with nearly 70 per cent of the city's churches being demolished. Hundreds of priests and bishops were arrested or executed, many becoming inmates of the Gulags.

Not until the German invasion of the USSR in 1941 did the Church enjoy any alleviation of Stalin's repressive policies—giving it a vital breathing space in which to bolster itself against the next onslaught. This came with

A religious procession in Kursk painted by Ilya Repin (1880–83). Although Repin's picture captures both the mass appeal of Russian Orthodoxy and its richly elaborate ceremonial, it is also a valuable social document laden with implied criticism. The distinction between the westernized gentry and the medieval conditions of the peasantry is highlighted both by the accurate depiction of their costumes and by the scene in the right background, in which a uniformed officer beats an offending serf.

the rise of Khrushchev who, though he proved willing to relax many of Stalin's repressive policies, remained determined to stamp out the Church's continued influence among the people.

Although active persecution lessened during the 1970s, it was not until the 1980s that Christians could practise their faith with anything resembling confidence and safety. Furthermore, church architecture and art were at last recognized as national treasures and a programme of renovation was initiated. Though limited in its effect, at least some of the devastation wrought during the earlier decades of Soviet rule was restored.

II: From Tatars to Time of Troubles

Rising phoenix-like from the ashes of Tatar subjugation, Muscovy came to dominate Russia under the powerful leadership of individuals such as Ivan III.

"They are called Tatars, or some say, Taurmens, or others, Pechenegs … We have written of them here for the sake of the memory of the princes of Rus and the misfortune that they brought upon them. For they made many countries captive, and slaughtered a multitude of the godless Polovtsy and drove others out."
First Novgorod Chronicle

The first Tatar foray into Rus occurred in 1223, and culminated in a Tatar defeat at the Kalka River. Undeterred, the Tatars undertook another reconnaissance in 1226, under Khan Batu, grandson of Genghis Khan. The Tatars returned in force in 1237 and destroyed Ryazan, Moscow and Vladimir. In 1240, when its elders refused to pay tribute, they destroyed Kiev. With the princes of Rus disunited, there was little resistance and Rus fell under the Tatar sphere of influence for two and a half centuries.

The Tatars saw their Great Khan as a divine power, mediating between heaven and earth, and thus all other nations were their subjects. They insisted on submission before negotiation; any nation refusing to submit was considered a rebel. They were divided into regional hordes; the Golden Horde, which dominated the Lower Volga, had its capital at Sarai. Their supreme authority was the Great Khan, while the Great Yasa (their code of laws) served to impose order and act as a unifying factor. Great emphasis was placed on the army, all male Tatars being drafted as soldiers. They reinforced their numbers by using the forces of conquered countries to help them in their further campaigns. As nomads, used to a mobile existence, they used their horse skills to enable them to develop a system of "post-horse" communication which facilitated control of large areas of territory by relatively small numbers.

The Tatars did not interfere directly in internal politics. Neither did they choose the actual prince but influenced that choice by giving or denying support to a particular prince and playing one off against another. But they did confirm the Grand Prince, who had to travel to Sarai to receive the *yarlik* entitling him to rule. Since the Great Yasa allowed religious toleration, all religions were respected and between 1240 and 1480 the Tatars showed remarkable tolerance towards the Russian Church.

In 1313 the official Tatar religion became Mohameddan yet tolerance towards the Russian Church continued. In 1270, the Khan imposed trading restrictions on Novgorod and then drove out the Yaroslav Grand Prince for refusing to comply. In support of the Tatars, Metropolitan Cyril threatened to excommunicate the princes if they did not obey. No taxes were levied on the Church and all their land was protected, even to the extent of issuing charters of immunity. Thus the Church became one of the largest landowners, land often being handed over to the Church for safe-keeping. This accumulation of great wealth explains why the Church became such a powerful political force.

Some historians have seen in the development of a centralized administration an imitation of the Tatar system. Taxation was clearly a hindrance to the existence of a free peasantry and this led ultimately to the development of serfdom. For the mass of the population Tatar rule meant an enormously increased burden in taxation and compulsory military service. For both purposes a general census covering the whole country was taken in 1257. In some areas direct rule was established, but it was more usual for local princes to be closely observed by officials. Princes had to produce taxes and troops for their Tatar overlords and this often sparked revolts which were suppressed with extreme savagery.

In 1238 Alexander, the son of Yaroslav, Grand Prince of Vladimir, became Prince of Novgorod. Desiring to convert Russia from Orthodoxy, and with the Swedes and the Teutonic Knights already in the process of converting the Baltic peoples, the Roman Church turned its attention toward Pskov and Novgorod. In 1240 Swedes invaded Russia at the mouth of the Neva. This invasion was defeated by Alexander whose victory was commemorated by bestowing on him the title "Nevsky". The Livonian Knights then moved against Pskov, but were defeated by Alexander in a battle on the frozen Lake Peipus.

Four years later Yaroslav died on his return from Mongolia. His sons, Andrei and Alexander, journeyed to the Tatar capital of Sarai to receive the *yarlik*, returning in 1249–50. Andrei became Grand Prince of Vladimir and Alexander Grand Prince of Kiev. In 1252 Nevsky went to the Golden Horde to muster armies against the Russian princes, first defeating Andrei who fled to Sweden; Nevsky was given the *yarlik* as Grand Prince, remaining a loyal servant to the Tatars. While Vladimir did not resist the census of 1257 there was strong popular feeling against it in Novgorod which culminated in rebellion. This was crushed and the census continued. Once again Nevsky went to the Horde to mediate on Novgorod's behalf and Novgorod was spared. In 1262 the northeast towns rebelled (over tribute and the burden of taxes). Nevsky went to the Horde yet again, persuading them not to destroy Russian towns, but he died on the way home.

Moscow, first mentioned in the *Chronicles* in 1147 when Prince Yurii Dolgorukii was responsible for its fortification, was destroyed by the Tatars in 1237. Its expansion commenced at the beginning of the 14th century; by 1304, at the end of the reign of Daniel—Moscow's first patrimonial prince and the younger son of Nevsky—most of the Oka and Volga area was under the control of Moscow. This period was also dominated by rivalry between Moscow and Tver. In 1304 the Grand Prince of Vladimir died and a succession crisis ensued. As there were no surviving sons, the throne should have gone to Mikhail of Tver, the eldest cousin, but the previous year Daniel of Moscow had died, leaving five sons, the eldest of whom, Yurii, contested Mikhail's rights to the throne of Vladimir. Since there was no precedent, laws were unwritten and tradition was insufficient to guarantee a legitimate heir, the final decision rested with the Khan of the Golden Horde. In 1304 both princes travelled to the Golden Horde and Mikhail received the *yarlik*. When he returned in 1305 he was placed on the throne of Vladimir, but faced opposition from Moscow, Pereyaslavl and Novgorod. In 1305 he marched on Moscow, defeated it, and a temporary peace was established between the two principalities. In 1306–08 the Tatars attempted for the first time to adjust the balance of power by supporting Yurii of Moscow, the principality which appeared the weakest. Most damaging to Mikhail was his failure to win Church support, especially the failure to secure his candidate as Metropolitan.

Between 1304–18 Mikhail held out against Yurii's intrigues though the latter eventually prevailed, by obtaining the support of the Russian Church, bestowing rich gifts on the Tatars and marrying the sister of Khan Uzbek. Eventually Uzbek revoked the *yarlik* given to Mikhail, removing him as Grand Prince. In 1318 Yurii was made Grand Prince of Vladimir and accompanied Tatar troops to oust Mikhail. Although Mikhail defeated Yurii, he was summoned to the Golden Horde and executed in 1319. Yurii was soon faced by resentment against increased taxes and general hostility from the Tver princes. When Yurii fell behind in his payments, the Tver princes complained to the Tatars; in 1322 the Tatars removed Yurii from the throne of

Vladimir. Mikhail's son, Dmitry, was then made Grand Prince. Yurii travelled to the Horde to complain, followed by Dmitry. Yurii was subsequently murdered there by Dmitry who, in turn, was executed by the Horde, installing Alexander, Dmitry's brother, as Grand Prince.

During the reign of Ivan I (Kalita—"Moneybags", reigned 1325–41), Moscow achieved supremacy. As a reward for suppressing a rebellion in Tver, Ivan, with the help of Tatar troops, was made Grand Prince of Vladimir. In 1380, Dmitry Donskoi defied the Tatars' demand for increased taxes and, leading an army representing an alliance of principalities, won a famous victory at Kulikovo Pole. But the victory was short lived; two years later, the Tatars returned in Dmitry's absence and sacked Moscow.

Moscow's success at Kulikovo can be explained by a number of factors. It was strategically placed in relation to Russia's river network, while to the east lay the thick forest belt which provided protection from outside attack. Tver could never equal Moscow either in wealth or troops and depended upon unreliable allies, such as Lithuania and the Tatars. Neither should the support from the Russian Church be underestimated; enormous land grants were made to the Metropolitan by the princes of Moscow; demonstrating its support, the Church moved its centre to Moscow. This was of paramount importance because the Church acted as a unifying and influential force in the country. Moscow princes shared a common purpose: to achieve power over all Russia; a strong desire for unity and, to this end, a determination to limit their progeny. Yurii, in Moscow, had no sons and his next three brothers died before him, with everything passing to Ivan II. Moscow's rival, Tver, had on average five or six sons, a factor encouraging de-centralization.

The Battle of Kulikovo, fought in 1380 near the River Don, greatly boosted Russian morale but did not prevent the Tatars returning in force and devastating Muscovy two years later. Overall, the wisest policy to follow in the face of Tatar military superiority was voluntary submission, as exemplified during the reign of Alexander Nevsky.

While the beginning of the 15th century was wracked with civil war, Muscovy was the creation of Ivan III (the "Great", 1462–1505). Although tribute was still being paid to the Tatars , this was of little importance since the Tatars were split and Moscow's major threat now came from the west: from Lithuania, the Livonian Knights and the Swedes. Adopting Tatar tactics of "divide and rule", in 1480 Ivan eventually overcame the Tatars and engaged in several campaigns against Lithuania.

Ivan wanted to become Grand Prince of "all Russia" ruling over a strong, independent and prestigious country. His domestic policy was dominated by this desire to replace the independent principalities with a centralized state. Tver, Riazan and Rostov all rebelled, while Novgorod, sandwiched between Moscow and Lithuania, resolutely refused to accept Ivan's supreme authority. In 1456 Vasily II led a punitive expedition to Novgorod and established a treaty limiting its powers, but its citizens refused to accept the conditions and called in Lithuania. A rebellion developed in 1470 and the pro-Lithuanian faction negotiated with the Uniate Metropolitan in Kiev. In 1471 Novgorod was routed, having failed to secure Polish help. Finally in 1477 Ivan attempted to eliminate Novgorod's independence by removing the *veche* and installing his own archbishop and others in important positions. Wishing to ensure all subjects were loyal when it came to tackling Lithuania, in the next decade he deported many wealthy residents (supposedly "disloyal" elements) and replaced them with his own servitors, granting them large estates. The final act of suppression came in the 1490s when Ivan threw out Hanseatic merchants.

With Novgorod crushed, the "gathering in of the Russian lands" continued; Tver submitted in 1485 and Viatka and Pskov in 1489 and 1510 respectively. To build a centralized state from the ruins of the appanage system required reform of the administrative system, of the system of land tenure and of the economy. On his death, Vasily had left appanages to his wife and his five brothers. Ivan, however, was determined to decrease their authority. In 1472 Yurii died and Ivan, to the annoyance of the other brothers, took control of his appanage. Ivan bought off his brothers with extra land and demanded that they obey him. In 1480, two brothers, Andrei and Boris, resenting his greed, rebelled against him. Ivan gradually reduced his brothers' powers until all the appanages had been annexed. He also imprisoned Andrei and confiscated his land.

In 1497 the *Sudebuik*, (or Unified Code of Laws) was issued providing one code of law for the entire country. This necessitated a huge expansion of bureaucracy. If a major war was to be waged against Lithuania then a large army would have to be organized with individuals personally responsible to Ivan. This led to the *pomeste* system, whereby a new class of service gentry, the *pomeshchiki* was created. By the terms of the *pomeste* system land was held in return for service and only service gave the right to land. On the death of a *pomeshchik*, the land passed to the Grand Prince. The service gentry became the nucleus for a permanent army as well as providing staff for a central administration. Such a scheme necessitated the urgent acquisition of land. Ivan, therefore, confiscated Andrei's lands, that of disloyal boyars, and of deported peoples. One result was a major conflict with the Church, the biggest land owner of the time. Though the Church was successful in this conflict, Ivan continued to undermine it by exploiting its divisions.

In 1439 the Russian Church, no longer accepting the superiority of the Patriarch of Constantinople and aware that Byzantium was in decline, asserted its independence. When the Byzantine emperor entered into relations with Rome, Russia viewed this as heresy and, in 1454, when Constantinople fell to the Turks, Russia viewed this as divine punishment. In 1470 Ivan III declared that the Patriarch of Constantinople had no control over Russia, and the belief emerged of Russia as the third Rome and Moscow as the centre of true Orthodox Christianity. The role of the Russian tsar, like that of the Byzantine emperor, was to protect the true faith—"a fourth Rome there will not be". Ivan and Vasily's deliberate cultivation of this idea can be seen in their adoption of the double-headed eagle and other Byzantine regalia, and in Ivan IV becoming the first Russian ruler to be crowned "tsar" (from caesar, earlier referring to the Byzantine emperor and also used by the Tatar Great Khan).

Ivan IV, the "Terrible" or "Dread" (*grozny*) (tsar 1533–84) continued territorial expansion as well as developing the centralized, autocratic Muscovite state inherited from Ivan III. The local aristocracy of boyars lost many privileges. The Muscovite annexation of surrounding states in the 16th and 17th centuries introduced many new boyar families into Moscow and this was resented by the old aristocracy. The new ones were considered socially superior to the old Moscow boyars, their place in society depending on the *mestnichestvo* system, where boyars were placed in an official social order determined by such factors as genealogy. Much boyar discontent came from loss of individual power and annexation of their territories. Conflict also arose over the attitudes adopted towards the west (Lithuania and Poland) by the ruling Muscovite family, many boyars disagreeing with the pro-Tatar and anti-Lithuanian/Polish stance. While dissatisfied, such opposition elements lacked unity.

After curbing the powers of the boyars in 1547, for 12 years Ivan ruled with the help of a personal council, which assisted him in implementing a programme of reforms in the 1550s and 1560s. These aimed at eradicating traditional freedoms, so that people would serve the tsar unconditionally, and the establishment of a unified system of service. The Law Code (*Sudebuik*) of 1550 restored laws which had earlier been weakened and both this and the Church Council were approved by the "Assembly of the Land", the *Zemskii Sobor*. A redistribution of lands was vital if a service state was to be erected while there were concerted attempts to recover lands lost during the chaotic years of his early reign. Hereditary land tenure (the *votchina* system) was abolished, tax immunities were discontinued and there was a renewed attack on Church immunity and ownership. Much of the land handed over to the Church for safe-keeping was confiscated and redistributed, Ivan having failed like Ivan III before him, in persuading the Church to give up estates voluntarily. A land census was introduced to enable verification of land ownership. During the 1550s there was evidence of a growth in land possessed by the service gentry. The logical culmination in 1555 was that every landowner had to provide service and this lasted until 1762 when the gentry were freed from service. To build up his administration, Ivan took over lands near Moscow and distributed them to his closest group, enabling them to support themselves but also making sure they were subject to him. Local government organizations (*zemstvos*) were introduced as tax gathering agencies and a list of peoples' tax assessments were compiled. To pay for this increased administration and foreign policy expenses, taxes were increased and the land census now enabled a new unit of tax to be imposed with discrimination against peasants and monasteries. There were also important developments in military reforms which realised the necessity for a regular army. A personal army (the *streltsy*) was established in 1550, also funded by levying a further tax.

During the 15th century, to reflect the growing prestige of the Muscovite state, the Moscow Kremlin was largely remodelled. Several new cathedrals were included in the design to indicate the power and influence of the Orthodox Church, the most impressive being the Uspensky and Arkhangelsky Cathedrals.

Ivan's foreign policy was aggressive, both against Lithuania and through the annexation of Tatar territories. Important military victories occurred between 1552–56 (the fall of the Tatar kingdoms of Kazan and Astrakhan) which opened the way to Siberia and the east. After Ivan fell seriously ill in 1553 he insisted that his councillors and other boyars pay allegiance to him and his son. Some refused and supported Vladimir, wanting to overthrow the monarchy. Ivan recovered but became more suspicious. There was also opposition to his declaration of war against Livonia in 1558, since his opponents desired war against the Tatars and the Turks. Ivan also blamed his opponents for the death of his first wife, Anastasia, in 1560, believing her to have been poisoned, rekindling unpleasant childhood memories of the probable poisoning of his mother when he was eight years old. Amidst crisis and talk of treason and plots, Ivan forced boyars to pay allegiance to him; others were executed. Matters worsened in 1564 with a military crisis in Lithuania and Ivan's fear of betrayal being reinforced when his friend Prince Kurbskii offered his services to the Lithuanians. Ivan left Moscow, blaming his departure on disloyal elements. Early in 1565 he established the *oprichnina*—his own special domain (with *oprichniki* or personal bodyguards)—to seek out treason and sedition. Even though the *oprichnina* included some boyars, the object of the *oprichniki* was to eliminate the boyars as a class. A reign of terror set in to prevent them from defecting and to ensure the tsar was autocrat over the entire country. Novgorod, retaining the last remnants of freedom and trade links with the west, was routed in 1570; its citizens were massacred and any survivors deported. Two-thirds of Novgorod was granted to the *oprichnina* which was finally abolished in 1572, Church privileges were to be eliminated and the metropolitan replaced.

When Ivan died in 1584 he was succeeded by his feeble son, Fyodor, though rule was effectively in the hands of Fyodor's brother-in-law, the regent Boris Godunov. After the death of Fyodor in 1589 and of Dmitry, Ivan's other son, in 1591—both in mysterious circumstances—Godunov was eventually elected tsar by the *Zemskii Sobor*, provoking discontent among the old boyar families. Even though he exiled the Romanovs with whom he was in conflict, Godunov could not command popular support. Amidst the epidemics, crop failures, the lawless bands that wondered the countryside, and the general discontent that beset 1601–03, rumours abounded that Dmitry was still alive and that Godunov was a usurper. A "Pretender", Dmitry, appeared in Poland and, though denounced by Godunov, was recognized by the Polish government as heir. With the help of Polish forces, Don Cossacks and some exiled boyars (all with grievances to air) this "False Dmitry" invaded Russia in 1604. Though he was defeated, when both Godunov and his son died, "False Dmitry" was proclaimed tsar. Dressed in western attire and with a Polish wife, Dmitry's widespread unpopularity led to boyar rebellion and his death. Chaos ensued and the "Time of Troubles" continued.

Poor harvests, famines and wars all acted to aggravate the already harsh conditions for the peasantry. The number of *pomeshchiki* increased and many more peasants were taken into the service structure, leading to widespread de-population of the central regions of the country. Many merchants also fled. To compensate, government measures became even more severe. And the lesser (weaker) *pomeshchiki* were hit the hardest, tending to lose their "service men". The government attempted to assert control over border areas especially in the Lower Volga but those who had fled from central Russia, formed free societies of adventurers with hetmen as chief, such as the Cossacks of the Lower Don. Russia built fortresses in these areas, which gradually developed into a frontier military region. Chaos ensued as areas supporting the first pretender rose up in rebellion. During a major peasant uprising under Bolotnikov, peasants ransacked gentry mansions and estates and marched on Moscow demanding the end of serfdom.

An 18th-century engraving shows envoys of Vasily III being received by the Holy Roman Emperor, Maximilian, in 1516. Under both Ivan the Great and Vasily, Muscovy continued a policy of expanding its diplomatic relations and embassies were sent to the Vatican and the Turkish Sultan as well as Maximilian.

Vasily Shuisky defeated them and eventually became tsar (1606–10) but had little popular support and was deposed by the *Zemskii Sobor*. For several years there was no accepted political authority with several Swedish and Polish claimants. The most successful pretender was the second "False Dmitry" who set up government outside Moscow. This second pretender, helped by Poland and Lithuania, was unsuccessful in his attempt to besiege Moscow but did manage to take the northern provinces. The Poles defeated Shuisky and marched on Moscow. In 1610–12 a Polish garrison held Moscow, elected a Polish Prince, Wladyslaw, and tried to reduce Russia to a Polish kingdom. The Patriarch opposed such attempts on the grounds that the people had no desire for Catholicism and denounced the Poles. In 1612 a popular militia, with Swedish assistance, drove out a mixed Polish and Russian force. A government was then set up in Yaroslav whose forces routed the Poles and took Moscow. In 1612, recognizing the need for national unity and the election of a tsar who had broad acceptance, the *Zemskii Sobor* elected a new tsar, the 16-year-old Mikhail Romanov, a relative of Ivan IV's popular first wife, with his father acting as Patriarch. Thus both the Church and the state were ruled by a Romanov, while the autocratic state and serfdom were consolidated. Three important lessons for Russia's future would emerge from "The Time of Troubles": the need for national unity; the danger of anarchy, and the threat from foreign intervention.

Russia under the Tatars

Mounting a series of increasingly daring and ambitious raids, the Tatars took advantage of Russian divisiveness.

The Tatars (or Mongols) who conquered Rus between 1219 to 1242 were predominantly Turkic. During an exploratory raid into the steppes in 1221, a strong force of Tatars swept out of the Caucasus, defeated a combined Rus and Polovtsian force at the Kalka River (1223), raided the Crimean and lower Dnieper area, moved northeast to the middle Volga—where they suffered one of their few defeats at the hands of the Volga Bulgars—then disappeared into Asia.

The Rus saw no more of the Tatars until 1237 when they returned in force under their commander, Batu, determined to subjugate the Rus and their neighbours to the west. First they overcame the Volga Bulgars, then set upon Vladimir-Suzdal, destroying its wealthy towns. Only the approach of spring saved Novgorod as the thaw would have trapped the invaders in the marshes. In 1239, the Rus princes failed to form a united front and the Tatars annihilated southwest Rus and sacked Kiev and hundreds of other settlements. By 1240, Rus resistance had virtually ceased.

The principality of Novgorod, protected by its geographical location, escaped invasion and submitted voluntarily to Tatar overlordship. But incessant attacks by Swedes and Germans weakened the principality. These attacks were repulsed by Novgorod's prince, Alexander Nevsky, who defeated the Swedes decisively on the Neva in 1240, and the Germans on the ice of Lake Peipus in 1242.

The Tatar invasion was to have lasting political, social and economic effects on Russia's subsequent development. The tribes of the "Golden Horde", Batu's descendants who were converted to Islam, were tolerant of other religions—allowing Russians to pursue Orthodoxy—and active in trade and agriculture. From their capital at Sarai, they held nine principalities in their power. Tribute was exacted from all inhabitants, irrespective of status and class and leading cities—centres of culture—were destroyed. Yet the Tatars also initiated a population census and influenced the Russian language as well as military (especially cavalry) development. The Tatars eventually withdrew to the steppes, restricting their intervention in the interior of Rus to punitive expeditions and the collection of taxes.

The Tatar warrior, Genghis Khan, in a drawing from a 13th-century Chinese manuscript. By the time of his death in 1227 the great Khan's empire stretched from the Black Sea to the Pacific coast.

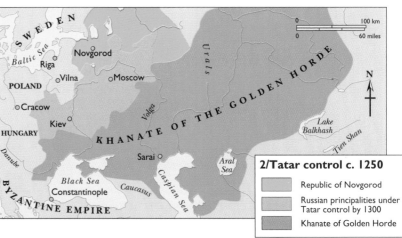

2/Tatar control c. 1250

Republic of Novgorod

Russian principalities under Tatar control by 1300

Khanate of Golden Horde

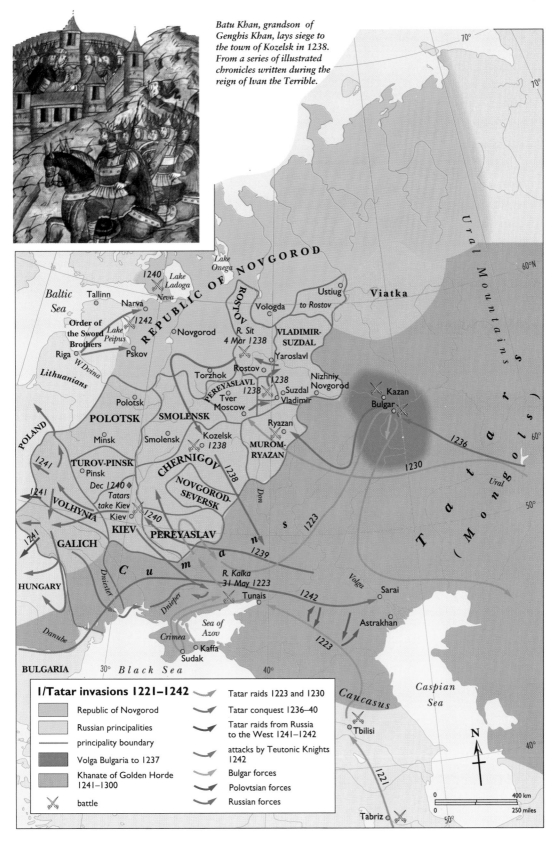

Batu Khan, grandson of Genghis Khan, lays siege to the town of Kozelsk in 1238. From a series of illustrated chronicles written during the reign of Ivan the Terrible.

I/Tatar invasions 1221–1242

- Republic of Novgorod
- Russian principalities
- principality boundary
- Volga Bulgaria to 1237
- Khanate of Golden Horde 1241–1300
- ✗ battle

- Tatar raids 1223 and 1230
- Tatar conquest 1236–40
- Tatar raids from Russia to the West 1241–1242
- attacks by Teutonic Knights 1242
- Bulgar forces
- Polovtsian forces
- Russian forces

Ural Mountains

Baltic Sea

Tallinn
Narva
1240 *Lake Ladoga* *Neva*
Lake Onega
1242
Order of the Sword Brothers *Lake Peipus*
Riga *W Dvina*
Lithuanians
Pskov
Novgorod
REPUBLIC OF NOVGOROD
ROSTOV
Vologda
Ustiug
to Rostov
Viatka

R. Sit 4 Mar 1238
VLADIMIR-SUZDAL
Yaroslavl
Rostov
Torzhok
1238
Nizhniy Novgorod
Kazan
1238
Suzdal
Vladimir
Bulgar
1236

PEREYASLAVL
Tver
Moscow
POLOTSK
Polotsk
SMOLENSK
Minsk
Smolensk
Kozelsk 1238
Ryazan
MUROM-RYAZAN
1238

POLAND
1241
TUROV-PINSK
Pinsk
CHERNIGOV
NOVGOROD-SEVERSK
1230
1238
Don

1241
Dec 1240 Tatars take Kiev
VOLHYNIA
Kiev
1240
KIEV
PEREYASLAV

Ural

1241
GALICH
HUNGARY
Dniester
C   u   m   a   n   s
1239
1223

R. Kalka 31 May 1223
Tunais
1242
Volga
Sarai
Astrakhan
1223

Dnieper
Sea of Azov
Crimea
Kaffa
Sudak

BULGARIA
30° *Black Sea*
40°
Caucasus
Caspian Sea

Tbilisi

N

Danube

1221

Tabriz
50°

70°
70°
60°N
60°
50°
40°

Tatars (Mongols)

0   400 km
0   250 miles

The Rise of Muscovy

"Let it be known to Your Majesty, O pious tsar, that all the realms of Orthodox Christendom have been reduced to your realm alone, and you alone on the earth bear the name of a Christian emperor."
Filofey, from his *Epistle to Vasily III*

As the Tatar stranglehold relaxed, the descendants of Alexander Nevsky followed a coherent policy of Muscovite expansion.

The period between 1261–1533 saw not only the end of Tatar domination but also the extension of Muscovy's enormous power. The origins of this astonishing rise in prestige and influence can perhaps be found in Alexander Nevsky's shrewd avoidance of the Tatar yoke during the mid-13th century. Accepting Tatar military supremacy, Alexander voluntarily submitted to their domination and paid tribute, thereby ensuring the succession of his son, Daniel, who became Prince of Moscow and founder of the Muscovite dynasty. It was to hold uninterrupted sway from 1276 to 1598.

During his reign Daniel succeeded in doubling his territorial possessions by expanding to the south, northeast and west. After a period of consolidation under Ivan I Kalita ("Moneybag"), further expansion took place under Dmitry Donskoi (reigned 1359–89) and Vasily I (reigned 1389–1425). During these reigns Muscovy was to grow to eight times its size by 1359. Not until the reign of Ivan III (reigned 1462–1505) was such impressive expansion to be repeated.

Under Ivan III Muscovy was finally relieved of Tatar domination—the Khan of the Great Horde finding himself incapable of enforcing payment of the tribute which Ivan had withheld for four years. Tatar raids continued, however, and the Muscovites were obliged to maintain a defensive barrier along the line of the River Oka. Ivan's territorial gains were impressive, swelling Muscovy to four times its size at his accession to the throne. Some minor princedoms were obtained through purchase, intimidation and voluntary submission, while others were conquered. Added to the acquisition of Yaroslavl, Rostov and Tver were extensive gains in the southwest and, most significantly, Novgorod and its northern colonies stretching even beyond the northern Urals. Campaigns between 1500–1503 against Lithuania in the west proved less effective, setting an ominous precedent for future conflicts in the region.

Ivan's expansionist policies were continued during the reign of his son, Vasily III who, besides taking Pskov and the remaining Ryazan lands, annexed Smolensk.

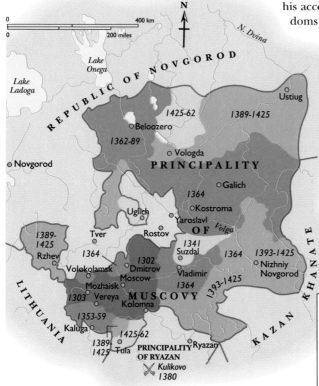

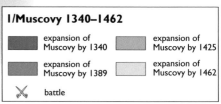

I/Muscovy 1340–1462

- expansion of Muscovy by 1340
- expansion of Muscovy by 1389
- expansion of Muscovy by 1425
- expansion of Muscovy by 1462
- ✗ battle

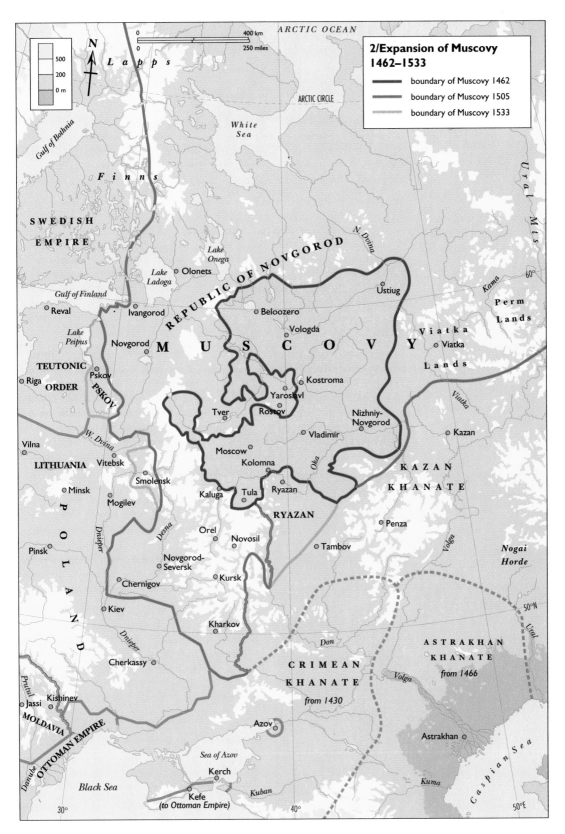

ARCTIC OCEAN

ARCTIC CIRCLE

2/Expansion of Muscovy 1462–1533

— boundary of Muscovy 1462
— boundary of Muscovy 1505
— boundary of Muscovy 1533

400 km
250 miles

N

L a p p s

F i n n s

White Sea

SWEDISH EMPIRE

Gulf of Bothnia

Lake Onega

Lake Ladoga

Olonets

Gulf of Finland

Reval

Ivangorod

REPUBLIC OF NOVGOROD

N. Dvina

Ustiug

Ural Mts

60°

Kama

Perm Lands

Lake Peipus

Novgorod

Beloozero

Vologda

M   U   S   C   O   V   Y

Viatka Lands

Viatka

TEUTONIC ORDER

Riga

Pskov

PSKOV

Kostroma

Yaroslavl

Viatka

W. Dvina

Tver

Rostov

Nizhniy-Novgorod

Kazan

Vilna

Vitebsk

Vladimir

LITHUANIA

Minsk

Moscow

Kolomna

Oka

KAZAN KHANATE

Mogilev

Smolensk

Kaluga

Tula

Ryazan

Desna

RYAZAN

Penza

Dnieper

Pinsk

Orel

Novosil

Tambov

Volga

Nogai Horde

P

Novgorod-Seversk

Kursk

50°N

Ural

Chernigov

O

L

Kiev

Kharkov

Don

ASTRAKHAN KHANATE
from 1466

A

Cherkassy

Dnieper

CRIMEAN

Volga

N

KHANATE

D

Prut

Kishinev

Jassi

from 1430

MOLDAVIA

OTTOMAN EMPIRE

Azov

Danube

Sea of Azov

Astrakhan

Kerch

Black Sea

Kefe
(to Ottoman Empire)

Kuban

Kuma

Caspian Sea

30°

40°

50°E

500
200
0 m

Russian Expansion

The reign of the idiosyncratic Ivan the Terrible was marked by successful expansion in the east and humiliating defeat in the west.

"We had thought
that you were the
monarch in your
own kingdom ...
Now we see that it
is really your people
who have the power
... In your virginal
state, you are just a
simple old maid.."
Ivan IV to
Elizabeth I of
England,
October 1570

The reign of Ivan IV (the "Terrible") was marked by spectacular successes in the east (symbolizing the turning of the Tatar–Moslem tide) and dismal failure in the west. In 1547 his coronation saw the first Muscovite ruler assume the title of tsar (caesar, emperor), indicating his determination to rule as an autocrat whose power was subject only to God.

In 1552 Ivan conquered the Tatar Kazan Khanate and in 1556 Russian forces subjugated the Astrakhan Khanate, bringing the entire Volga basin under Ivan's control. A year later he subsumed the Kabarda princes (northern Caucasus) and the Great Nogai Horde (east of the lower Volga) and—at about the same time—the Circassians (who inhabited the region south of the Kuban River). In the sense that he ruled a multinational empire, Ivan became the successor to the Golden Horde. Further eastward, expansion came through private initiative in about 1581 when the Cossack Yermak undertook the conquest of western Siberia, which was nominally completed by about 1584.

In contrast to such eastern triumphs, efforts at expansion to the west were far less successful. The Russian invasion of Livonia (approximately modern Latvia and Estonia) was soon to develop into a war with Sweden, Poland and Lithuania, lasting until 1583. Poland absorbed Lithuania (in 1569) and the treaty with Poland in 1582 cost Ivan the gains in Livonia he had won earlier. The peace with Sweden in 1585, moreover, cost him all the Russian possessions on the Gulf of Finland, as well as the western shore of Lake Ladoga (including Kexholm, the Swedish name for Karelia). The accidental discovery by English merchant explorers in 1553 of the route to Russia via the White Sea partially compensated for the loss of outlets on the Baltic. In 1584 the port of Archangel was founded at the mouth of the Northern Dvina River. The only territorial gain in the west which was retained during Ivan's reign was the Sebezh area (northwest of Polotsk), captured from Lithuania in 1535 when Ivan was a child.

After Ivan's death in 1584 power was exercised by Boris Godunov, brother-in-law to Ivan's feeble and ineffectual son, Fyodor I. Godunov continued the penetration of Siberia and waged successful war in the northwest, recovering lands on the eastern Baltic coast which Ivan had previously lost to Sweden during the Livonian War.

Ivan the Terrible, whose name became a byword for tyranny and savage, unpredictable cruelty. Though beginning his reign auspiciously, Ivan's dark side was brought to the fore by the death of his first wife Anastasia—whom the Tsar believed to have been poisoned. While guilty of such atrocities as the sacking of Novgorod, Ivan also had less vicious propensities, seeking recreation in the composition of liturgical music.

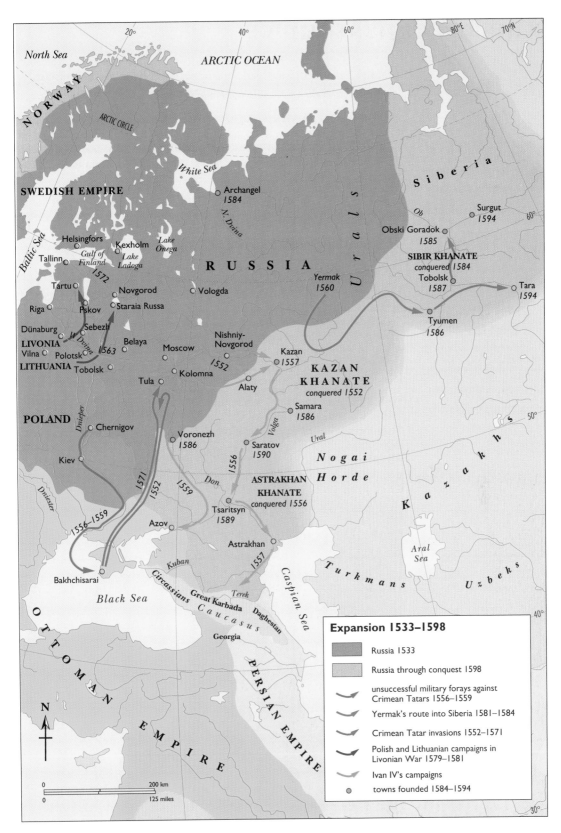

North Sea

ARCTIC OCEAN

20°   40°   60°   80°E   70°N

NORWAY

ARCTIC CIRCLE

White Sea

Siberia

SWEDISH EMPIRE

Archangel
1584

N. Dvina

Ob

Surgut
1594

Obski Goradok
1585

60°

Helsingfors

Kexholm

Lake
Onega

RUSSIA

Urals

Yermak
1560

SIBIR KHANATE
conquered 1584

Tobolsk
1587

Tara
1594

Baltic Sea

Tallinn

Gulf of
Finland

Lake
Ladoga

1572

Tartu

Novgorod

Vologda

Tyumen
1586

Riga

Pskov

Staraia Russa

Dünaburg

Sebezh

W. Dvina

Nishniy-
Novgorod

LIVONIA

Belaya

Moscow

Kazan
1557

**KAZAN
KHANATE**
conquered 1552

Vilna

Polotsk

1563

1552

LITHUANIA   Tobolsk

Kolomna

Tula

Alaty

Samara
1586

50°

POLAND

Chernigov

Dnieper

Voronezh
1586

Saratov
1590

Volga

Ural

N o g a i

K a z a k h s

Kiev

1571

1552

1559

Don

Dnister

1556

**ASTRAKHAN
KHANATE**
conquered 1556

H o r d e

1556–1559

Azov

Tsaritsyn
1589

Astrakhan

Aral
Sea

40°

Kuban

1557

Caspian Sea

T u r k m a n s

U z b e k s

Bakhchisarai

Black Sea

Circassians

Great Karbada

Terek

Caucasus

Daghestan

Georgia

Expansion 1533–1598

▨	Russia 1533
▨	Russia through conquest 1598
↝	unsuccessful military forays against Crimean Tatars 1556–1559
↝	Yermak's route into Siberia 1581–1584
↝	Crimean Tatar invasions 1552–1571
↝	Polish and Lithuanian campaigns in Livonian War 1579–1581
↝	Ivan IV's campaigns
●	towns founded 1584–1594

**O T T O M A N
E M P I R E**

P E R S I A N E M P I R E

N

30°

0   200 km
0   125 miles

The Time of Troubles

Rocked by internal discord Russia suffered from the incursions of its western neighbours intent upon territorial expansion.

"[False Dmitry] was always accompanied by a large armed escort. Before him and behind him marched armed soldiers bristling with pikes and halberds … This mass of glinting weaponry made a terrifying spectacle."

Avraamy Palitsyn

After the end of the "Rurik" dynasty in 1598, Russia provided opportunities for its western enemies, both the Poles and the Swedes seizing important western areas (the former even briefly holding Moscow) during the "troubled" years between the death of Fyodor I and the coronation of the first Romanov tsar, Mikhail in 1613.

Among the numerous complex campaigns of the period, the following had enduring significance: Smolensk fell to the Poles (after a 20-month siege); Novgorod accepted Swedish overlordship in 1611; by 1613 the Swedes controlled the northwest, from Lake Ladoga south to Lake Ilmen and west to the Gulf of Finland; by 1612, the Poles had occupied an even larger area, including Moscow itself.

Eventually, the Russians defeated attempts to impose foreign domination on the country. By the terms of the Peace of Stolbovo in 1617, Sweden withdrew from Novgorod but retained most of the Lake Ladoga area (including Kexholm), the territory north of the lake, and Ingria. Since the Swedes had acquired Estonia in the 16th century, the addition of Ingria gave them control of the entire littoral of the Gulf of Finland. Russia was again cut off from the West via the Baltic Sea.

With the Polish army once again advancing on Moscow in 1618, the Polish threat was more ominous. By the terms of the Deulino Armistice (1618), Poland retained the provinces of Smolensk and Seversk, and a strip of varying width along most of Russia's western border. Yet no one party was prepared to accept such settlements as final, and inconclusive foreign intervention during the "Time of Troubles" set the scene for future wars.

Ambassadors of the Moscow Zemstvo entreat Mikhail Fyodorovitch to accept the Tsar's crown in 1613. Mikhail was a relative of the popular first wife of Ivan the Terrible and his accession to the throne marked the end of the Time of Troubles.

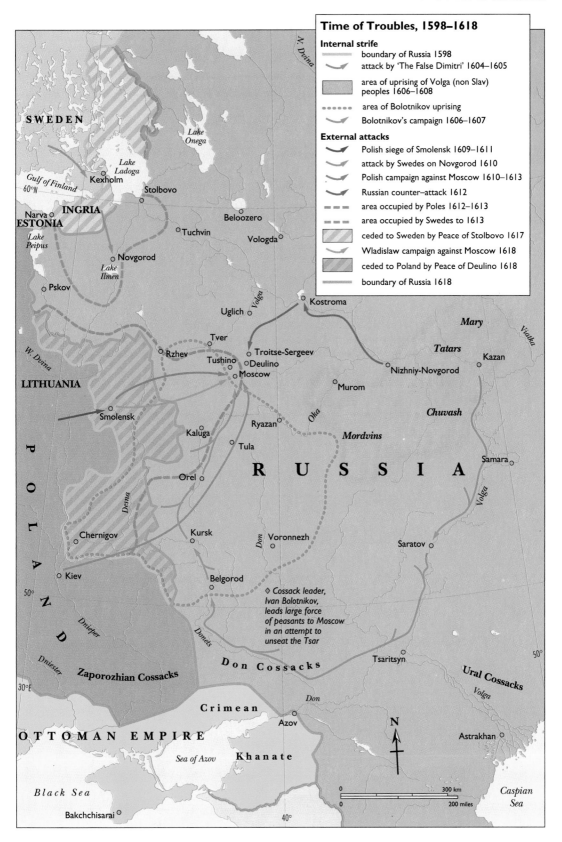

Time of Troubles, 1598–1618

Internal strife

- boundary of Russia 1598
- attack by 'The False Dimitri' 1604–1605
- area of uprising of Volga (non Slav) peoples 1606–1608
- area of Bolotnikov uprising
- Bolotnikov's campaign 1606–1607

External attacks

- Polish siege of Smolensk 1609–1611
- attack by Swedes on Novgorod 1610
- Polish campaign against Moscow 1610–1613
- Russian counter–attack 1612
- area occupied by Poles 1612–1613
- area occupied by Swedes to 1613
- ceded to Sweden by Peace of Stolbovo 1617
- Wladislaw campaign against Moscow 1618
- ceded to Poland by Peace of Deulino 1618
- boundary of Russia 1618

SWEDEN

Lake Onega

Lake Ladoga

Gulf of Finland
60°N

Kexholm

Stolbovo

INGRIA

Narva
ESTONIA

Lake Peipus

Tuchvin

Beloozero

Vologda

Novgorod

Lake Ilmen

Pskov

Kostroma

Uglich

Tver

Mary

Tatars

Viatka

Kazan

Rzhev

Troitse-Sergeev

Tushino   Deulino

Moscow

Nizhniy-Novgorod

LITHUANIA

Murom

Smolensk

Ryazan

Oka

Chuvash

P
O
L
A
N
D

Kaluga

Tula

Mordvins

Samara

Orel

R U S S I A

Volga

Chernigov

Kursk

Don

Voronnezh

Saratov

Kiev
50°

Belgorod

◇ Cossack leader, Ivan Bolotnikov, leads large force of peasants to Moscow in an attempt to unseat the Tsar

Tsaritsyn

Ural Cossacks

Volga

50°

Dnieper

Donets

Don Cossacks

Dniester
30°E

Zaporozhian Cossacks

Crimean

Don

N

Astrakhan

OTTOMAN EMPIRE

Azov

Khanate

Sea of Azov

Black Sea

Bakchchisarai

40°

0   300 km
0   200 miles

Caspian Sea

W. Dvina

V. Dvina

Desna

The Fur Trade

Spurred on by private mercantile initiative, Russian explorers and trappers pushed further and further eastwards in search of highly-prized pelts.

During the 17th century Muscovy made spectacular advances eastwards. Having conquered Kazan and Astrakhan in the 1550s, Russians then advanced across the Urals, and on through the river systems that intersect the huge Siberian plain, to Lake Baikal. Some expeditions were planned by the state. Others—after the tsar had sanctioned such operations into Siberia—by extremely wealthy merchant families, such as the Demidovs and Stroganovs. In 1581–82, the latter family employed a group of Cossacks under Yermak and captured the western Siberian Khan's citadel of Isker, on the Irtysh River. From there they went on to establish the new Siberian capital of Tobolsk. Thus the way was open for explorers, seeking fur tribute (or *yasak*) from the indigenous peoples, to push ever onwards.

This 16th-century woodcut by Michael Peterle shows a procession of Russian ambassadors with their train of fur-bearing merchants. The state did not fund its emissaries and they were expected to live on the proceeds of trade, which was based predominantly, though not exclusively, on the sale of highly valued furs.

The motive for this eastwards expansion was economic; a search for luxury furs or "soft gold", which in earlier times were considered the mainstay of the Russian economy. Furs were a valuable trading commodity, and fur-bearing animals were ruthlessly hunted. By the 17th century, stocks had been depleted in the northern regions of European Russia. Using the rivers, the hunters pushed eastwards across the Urals to tap the stocks of sable and squirrel in Siberia. This led to clashes between Russians and the natives whose stocks were being slaughtered. Following the founding of Irkutsk in

1632, in the 1640s explorers penetrated mountainous territory to the Amur River basin and the Pacific Ocean. (Semyon Dezhnev, another Cossack adventurer, became the first person known to have sailed up-river to the Arctic Ocean and from thence round the tip of Siberia through what was later named the Bering Straits.) In the eastern ocean, hunting changed to the sea otter (or sea beaver as they were called) and fur seal of the North Pacific. Prices for sea otter skins in North China exceeded those of sable pelts in western Europe. In Alaska, the Russians pursued the sea otter almost to extinction. Eventually the fur trade went into decline, partly due to lack of sources and the collapse of the market.

Subjugated during the second half of the 16th century, the Khanate of Sibir (Siberia) proved to be a treasure house for merchant families such as the Stroganovs. In quest of valuable furs, the merchant explorers advanced rapidly, making full use of the region's broad navigable rivers.

III: From Muscovite to Imperial Russia

Having stabilized after the tumultuous Time of Troubles, Russia continued to be dominated by its autocratic tsars, under whose influence it gradually turned to the West.

"The metropolitans and the archbishops and all conditions of men came to Moscow from all the towns and monasteries and they set about choosing a sovereign. ... they all cried out with a loud voice that they desired Mikhail Fyodorovich to reign over Muscovy."
Nikon Chronicle

With the establishment of the Romanov dynasty in 1613 under Mikhail Romanov (1613–45) and his son, Alexis Mikhailovich (1645–76), there was more than half a century of stability. After debilitating wars in the 1650s and 1660s, a modest prosperity was achieved, exemplified by buildings from the period. Internal stability also facilitated territorial expansion, with the successful incorporation into Muscovy of large portions of the regions now known as Ukraine and Belarus. Although the western European nations remained largely dismissive of Russia, the indifference was not mutual. Peter's immediate predecessors, Alexis, Fyodor and the regent Sophia, all exhibited a desire for broadening Russia's cultural horizons, and from 1652 westerners were granted their own suburb within the precincts of Moscow. Reluctant though they may have been, the rulers of Muscovy were forced to accept that their autocratic methods were becoming increasingly anachronistic. Their hesitant acceptance of western influence was to culminate in the energetic and uncompromising westernizing policies of Peter the Great.

Changes in the western boundary were prompted by the new Romanov dynasty seeking revenge for the earlier Polish invasions (an attempt to reacquire the western lands of Kievan Rus) although the attempt to liberate Smolensk (1632) was a failure and the subsequent Peace of Polyanovka (1634) brought only minor border changes. In 1648 unrest in Ukraine turned into warfare between the Zaporozhian Cossacks (led by Bogdan Khmelnitsky) and the Poles, leading eventually to another war between Poland and Russia (1654–57) after Tsar Alexis took the Cossacks under his protection. Ukraine and Poland were both devastated as a result of the constant fighting during these years. Smolensk, Chernigov, the entire left bank of Ukraine (east of the Dnieper River) and Kiev on the right bank were all ceded to Russia by a weakened Poland by the Armistice of Andrusovo, 1667. In 1678 Russia ceded to Poland a small area (in the northwest, including the towns of Sebezh and Nevel) in exchange for Kiev—all these Russian "gains" being confirmed by the Eternal Peace of 1686. Conflict among the Cossacks had meanwhile resulted in their division, roughly along the line of the Dnieper, into separate rival groups.

Russian expansion also continued in the east. By 1600 peace had been reached with the west Siberian Khanate of Kuchum and this removed the final barrier to expansion to the Pacific Ocean. Russian fur traders then continued to push the frontier ever further eastwards. Cossacks exacted tribute in furs from natives and set up small garrisons along the river routes. The outpost of Yakutsk was established in the middle Lena River valley in 1632 and among other places Okhotsk, on the northern shore of the sea of that name, was founded some 17 years later. By 1689 the entire territory to the Pacific, excluding Kamchatka, was in Russian hands, even though frequent native uprisings (especially in the extreme northeast) revealed the relatively ineffective control of the small and frequently disputant Cossack garrisons. Expansion had simultaneously occurred along the southern fringes; the founding of Guryev near the mouth of the Ural brought the river's right bank under Russian control in about 1645. Further to the east, Irkutsk was founded near the southwestern end of Lake Baikal 7 years later.

Expansion, especially around the Amur River basin, was effectively checked. The region was first explored (from Yakutsk) around 1643 but Russian aggression soon led to native appeals for assistance to the Manchu Emperor. The fort at Albazin on the upper Amur, was witness to numerous bloody encounters for almost 30 years, largely because the cannon used by the Chinese constituted the most sophisticated resistance to the Cossacks throughout the whole of Siberia. Eventually, skilled Chinese diplomacy succeeded in securing a peace treaty, agreed at Merchinsk in 1689, which established the boundary along the watershed of the Stanovoy Mountains, effectively excluding Russia from both banks of the Amur and from its tributaries east of the Argun River. This treaty was to last for 150 years.

Russia's first great popular rebellion was led by Stepan (Stenka) Razin during the reign of Tsar Alexis. Razin was a Don Cossack who, with a band of followers, engaged in daring exploits as a pirate and freebooter on the lower Volga and along the Caspian to Persia. Early in 1670 his flotilla sailed up the Volga, stopping to slaughter officials and landlords and to proclaim liberation. His following grew alarmingly until government forces defeated him at Simbirsk. Razin's base at Astrakhan held out for a time but he was captured and subsequently executed in Moscow. His name, and that of his supposed son, lived on in countless folk legends and songs as a deliverer who would bring freedom and justice.

The reign of Peter I (the "Great" 1682–1725), saw almost constant warfare culminating in a victory over Sweden (with permanent and significant annexations in the northwest), though only a stalemate with Turkey and transitory gains in Persia. Peter's early campaigns against the Turks and their Crimean Tatar vassals pushed down the Dnieper and Don valleys—the former leading to several forts being established which extended Russian influence over the Cossacks—until Azov was captured (1696) and a new naval base established at Taganrog on the Sea of Azov. These gains were confirmed in the Treaty of Constantinople in 1700, but were soon lost after Peter's forces were overpowered by a Turkish–Tatar force at the Prut River in 1711. Russia was forced to abandon Azov, Taganrog, the new fleet and the Dnieper River forts (confirmed in the Peace of Adrianople in 1713). While Turkish control over the Zaporozhian Cossacks increased, the Ottoman Empire was henceforth on the defensive against Russia. In 1716 Omsk was founded on the middle Irtysh River and two years later, Semipalatinsk, further upstream. Kamchatka was taken (after its discovery or re-discovery in 1697) and the Kurile Islands were explored and claimed for Russia by Cossacks sometime around 1711.

Following the Great Northern War with Sweden (1700–21), Peter conquered Ingria (by 1703) and founded St. Petersburg, Russia's new capital, a decade later. In the Treaty of Nystadt (1721) Russia annexed Estonia, Livonia, the Dago and Osel Islands, together with large areas ceded in southeastern Finland, including the provinces of Kexholm and Vyborg, pushing the new boundary farther west along the northern shore of the Gulf of Finland. No Russian territory in future was to be annexed by Sweden. Peter's adoption of the title "Emperor" was symbolic of Russia's new stature as a European power. In the last years of his reign, Russia's invasion of Persia (1722–23), culminated in the annexation of the southern and western shores of the Caspian Sea—including the towns of Derbent and Baku—though a few years after Peter's death Russia withdrew from these areas to ensure Persian assistance in Russia's war with Turkey.

While much of his rule was taken up by wars and territorial acquisition, Peter's reign was also known for its domestic reform and westernization.

Peter appears to have had a vision of a properly ordered society, linked to military objectives and political and economic needs. The old boyar Duma (parliament) and *Zemskii Sobor* (assembly of the land) were replaced by a ruling senate. Government departments were amalgamated and a foreign expert was included in the committee of each newly-created ministry. Church administration was also revised, with the creation of a governing synod, chaired by a layman.

To pay for his wars and military campaigns, Peter needed to raise revenue, and much of it came from the poll or soul tax which fell most heavily on the peasants (Peter eventually taxed anything; in an attempt to westernize his boyars he even taxed beards). All members of society were expected to serve the state according to their condition, providing a system through which anyone, theoretically, could progress. There was limited growth in education; arranged marriages ceased for noble women; and serfdom increased, extending into industry.

While the organization of government bodies, rank and titles, education syllabuses, town names, painting and architectural styles all had a distinctly foreign or "Western" feel, these innovations were also anti-Muscovite. Peter appreciated the need for trade and military success and his interest in commerce and industry and natural sciences was more indicative of western "progress" than what was typically Russian. Change though was limited. While there was great access to information, strict control was placed over divergence of views. For Peter, it seems, propaganda was more important than the free flow of knowledge. Foreign experts were welcomed, attracted by large salaries and opportunities. The service state itself, underpinned by Peter's absolutism, was hardly progressive. Yet Peter succeeded in gaining the respect of non-Russians and their positive impressions were strengthened by the Tsar's conscientiousness and his willingness to converse with even the most humble of his subjects.

While weaker than Peter, his successors (Elizabeth, Peter III, Anne and Catherine II) achieved greater success against the Turks. After four years of war, in 1739 Russia regained Azov together with both sides of the Don River estuary, even though Russia was prohibited from fortifying the area and from using the Azov and Black Seas for shipping. The most strategic acquisition was an area of the steppes between the Southern Bug and Don Rivers just north of the Crimean Tatar domain which included all of Zaporozhie. In an attempt to increase Russian control of the area, two regiments of Serbs were settled on the right bank of the Dnieper River in 1752 (giving the area its short-lived name, New Serbia). In later years other Serbs were settled between the Dnieper and Donets Rivers, east of Zaporozhie. During the reign of Anne (1730–40), nearly all of the Caspian littoral acquired by Peter was voluntarily ceded to Persia.

In the west another gain came from the Treaty of Abo (or Turku) in 1743, concluding a war initiated by Sweden in an attempt to recoup her losses of 1721. While Cossack forces had occupied much of southern Finland, Russia annexed a small area in the southeast, just west and northwest of Vyborg, producing a frontier along the Gulf of Finland following the Kymi River.

In Asia, nearly all of Russia's earlier annexations along the Caspian had been restored to Persia by 1735: all lands south of the Kura River in 1732 and the boundary moved north to the Sulak River in 1735. Yet elsewhere native resistance had been checked by 1732, and in 1740 the naval base of Petropavlovsk was founded on the southeastern coast.

In 1731 the Kazakhs of the Lesser (Younger) Horde nominally accepted

Russian sovereignty, extending Russian influence almost to the Aral Sea. In *c.* 1740 such suzerainty was also accepted by part of the Middle Horde, though firm Russian control over these Kazakhs had to wait until the 19th century.

North of Kazakh territory further permanent gains came with the founding of Orsk (*c.* 1735) and Orenburg (*c.* 1743) on the Ural River, with the latter fort serving as a base for subsequent expeditions deep into Central Asia. Another fort—Iletskaya Zashita—was established in 1743 and the town of Petropavlovsk on the Ishim River in 1752.

It is also possible that Alaska may have been sighted first by Russians in 1732, the year when Fedorov and Gvozdev are said to have reached Cape Prince of Wales on the eastern shore of the Bering Strait. Other expeditions followed: Bering reached North America in 1741 and the first of many merchant voyages there (to hunt for furs) was undertaken two years later. Exploration of the Aleutian Islands (discovered 1741) also occurred at this time. More ambitious Russian plans for North Amercia were to come in the following century.

The fortress of Schlüsselburg near St. Petersburg, in a late 18th-century engraving. It was in this formidable prison that Empress Anne's heir, Ivan VI, met his death. Confined by the adherents of Elizabeth, Ivan was murdered in 1764 after a failed attempt to proclaim him emperor.

Catherine II ("the Great', reigned 1762–96) was suddenly catapulted to the throne after the *coup d'état* against her husband, Peter III. During the first ten years of her rule, prolonged war with Turkey exerted mounting pressures of taxation and recruitment on the already burdened serfs (whose protests marked each year of her reign), while the liberation of the gentry in 1762 raised peasant hopes of similar freedom. The Pugachev rebellion, led by a Don Cossack, Emelyan Pugachev, originated in the southern frontier in 1773 and reverberated throughout the Volga and Urals regions as a series of local uprisings. Most of Pugachev's recruits were peasants or lesser

Cossacks, lower clergy, traders and craftsmen from the towns, Volga boatmen and Urals foundry workers, though nomads, mountain tribesmen and religious dissenters were also included.

While the main concern of the rebellion was the encroachment of central government in the freedom of individual groups, and the threat this posed to their independence, peasant issues did feature. The rising was directed mainly against the gentry—freedom from landlord control being a key demand—and the bureaucracy and the state it administered rather than the monarch personally. Religious and social myths played a key role in the spread of rebellion and Pugachev personally appreciated the power of propaganda. The patent illegality of Catherine's accession and Peter III's abrupt and mysterious death were particularly apposite material, while even the sex of the new sovereign may have undermined her legitimacy in the minds of the patriarchal peasantry.

Sowing seeds of doubt about Catherine's right to rule the empire, Pugachev claimed that he himself was the "true tsar" (Peter III) who was the "protector of the people". As pretender to the throne he was able to air a wide range of political, economic, social and religious grievances and thus provide a banner of legality for peasant rebellion.

Pugachev's proposals encompassed grievances of various groups of supporters and offered an alternative form of politics. His seditious letters contained promises to the Yaik Cossacks that traditional liberties, such as fishing and hunting rights (the undermining of which had already led to open revolt) would be restored; promises of freedom for the Bashkirs; promises of redress for the Old Believers (religious dissidents who had broken away from the Church in the 17th century in protest against reforms); and promises of liberation to the peasant workers who were labouring in the Ural factories and mines.

Grigori Potemkin, painted by Giovanni-Baltista Campi (c.1790). Created Prince of Tauris by Catherine, Potemkin encouraged his Empress in grandiose foreign-policy schemes, chief among which was his "Greek Project" which involved nothing less then a revival of the Byzantine Empire.

Pugachev's aim was to replace Catherine's corrupt government with a Cossack-style democracy, which would involve the abolition of serfdom and forced labour in industry—in essence, the people of Russia would become servants of the tsar. However, the rebellion failed to undermine the fabric of the tsarist state. Lack of a coherent programme was aggravated by serious rivalries between different groups and the unchallenged superiority of government forces. Pugachev was betrayed to the authorities, captured and executed. His rebellion was suppressed in 1774 with much bloodshed.

Catherine's reign was characterized by aggression and annexation. The Crimean Tatars were subjugated, the Polish nation crushed, the conquest of the Caucasus begun, the Aleutian Islands annexed and the first Russian settlement in Alaska authorized.

The most important acquisitions were from Poland. The first partition in 1772—jointly with Prussia and Austria—saw annexation of the Belorussian areas of Polotsk, Vitebsk and Mogilev, together with part of Lithuania, giving Russia the right bank of the western Dvina River and the left bank of the Dnieper. The second partition in 1793, together with Prussia, saw Russia acquire Minsk (Belorussia) and right bank Ukraine (Ukraine west of the Dnieper). While ethnic and historical factors (the areas had been part of Kievan Rus) could be used to justify these partitions, the same could not be said for the third partition in 1795; again involving all three countries as in the first, this saw the acquisition of the remaining parts of Belarus (Courland—which it can be argued voluntarily accepted Russian rule—and the rest of Lithuania) and areas which had never before been under

Russian control. For the first time in its history, Russia had common borders with Prussia and Austria, its western frontier following stretches of the Niemen, Western Bug and Dniester rivers.

Two victories over Turkey saw the annexation of most of the northern shore of the Black Sea. The Treaty of Kuchuk-Kainardzhi (1774) provided Russia with a foothold directly on the coast, between the Southern Bug and Dniester Rivers, (including the port of Kherson), while the 1739 restrictions were lifted with regard to fortification of the Azov area and significant political and economic rights were acquired within the Ottoman Empire. The Crimea was declared independent from Turkey and the ports of Kerch and Enikale were annexed. The Kabarda area of northern Caucasus was granted to Russia, although real control only came in the 19th century.

Crimean Tatar "independence" in 1774 was followed—under strong Russian pressure—by the formal annexation of their state in 1783. Russia now possessed both sides of the Sea of Azov, and south to the Kuban River, thus ending centuries of conflict with the Crimean descendants of Batu's far-ranging Tatars.

Following a new Turkish war, the Treaty of Jassy in 1791 (1792 in the new western calendar), confirmed Russia's annexation of the Crimea, granted Russia the northwestern Black Sea littoral between the Southern Bug and the Dniester Rivers (including the ports of Ochakov and Odessa). Catherine had succeeded where Peter had failed: Russia was now a Black Sea power with a fleet to prove it. Some 200,000 square miles of territory were acquired under Catherine although the inhabitants by and large remained hostile to Russian rule.

Catherine's reign saw great increase in educational levels and opportunities, including the education of women. Books were also published in some number, Catherine herself promoting satirical journals and writing plays. It was in her reign that the founders of modern Russian literature and public opinion began to emerge including such figures as the playwright Fonvizin, the poet Derzhavin, the poet and prose writer Karamzin, and probably the best known of all, Alexander Radishchev, for his biting attack on serfdom in his *Journey from St. Petersburg to Moscow*. Catherine took a particular dislike to it because it warned of popular rebellion and Radishchev was eventually exiled, his death sentence for treason having been commuted. Novikov, the publicist and publisher, was also arrested, though his detention was probably owing to his membership of the Freemasons—an organisation which Catherine viewed with deep suspicion. Although Catherine was anxious to be seen as an enlightened monarch it is also clear that she would countenance no relaxation of her autocratic powers. Theatre, music, painting, architecture and building all flourished during her reign.

A jewel-encrusted cup presented to Peter the Great by his son Alexis in 1694. Despite the exchange of such rich gifts the relationship between father and son was extremely strained. The powerful and pugnacious Peter despised his meek and aesthetic heir and their continued differences terminated in Alexis's execution for treason.

Russia to the End of the Petrine Period

Despite the initial reluctance of the first Romanov tsars, Russia extended its boundaries and under Peter I moved to the west.

"... the tsar is mechanically inclined, and seems designed by nature to be a ship's carpenter rather than a great prince."
Gilbert Burnet, Bishop of Salisbury

The period between the Time of Troubles (1598–1618) and the accession of Peter I (the Great) in 1682, was one of consolidation and gradual expansion. Tsar Mikhail (tsar 1613–45) established civic order and achieved agreements with the Swedes and Poles, but failed to regain Smolensk from the latter. He also drew back from confrontations with the Ottoman Empire when in 1642 the Don Cossacks offered him the Turkish fortress of Azov, which they had temporarily seized. Likewise, Tsar Alexis (tsar 1645–76) initially proved unwilling to incorporate Ukraine into his territories, despite offers of allegiance from the Russian Orthodox Ukrainian leader, Bogdan Khmelnitsky. When a union was eventually agreed in 1654, the subsequent war with Poland favoured the Muscovites who, under the Armistice of Andrusovo of 1667, acquired Kiev and Ukraine on the left bank of the Dnieper. The penetration of Siberia was also completed, extending Muscovite influence to the Sea of Okhotsk. During this period, possibly the most traumatic upheaval was the rebellion in 1670–71 of the Don Cossack, Stenka Razin, which, prior to its suppression by the Tsar's army, rocked the region of the lower and middle Volga.

Russian adherence to the anti-Ottoman Holy League involved Muscovy in war with the Turks and two unsuccessful campaigns against the latter's vassals, the Crimean Tatars, in 1686 and 1687. Peter the Great (tsar 1682–1725), continued the war with an unsuccessful expedition against Azov in 1695. Undeterred, Peter returned the following year at the head of a newly-built fleet which successfully subdued the Turkish garrison; however, the fortress was lost again in 1713. Though he began his tenure with a campaign in the south, Peter's overwhelming obsession was to modernize and westernize his country, and this aim was reflected as much in his territorial policies as in his political and social reforms. Intent upon breaking through to western Europe, Peter embarked upon the long, costly—but ultimately successful—war with Sweden which began in 1700 and ended in 1721 when the Peace of Nystadt was signed. Through success in the Northern War, ,Russia gained Ingria and Karelia and took possession of the entire Baltic coastline down to Riga. Modernized and developed under Peter, Muscovy (known from 1721 as the Russian Empire), was prepared for further expansion under its 18th-century empresses.

Peter the Great at the Battle of Poltava in June 1709. One of the most crucial battles in European history, Poltava ended in victory for Peter's reformed and restructured army. Peace between Russia and Sweden was finally signed at Nystadt in 1721.

2/ Eastward expansion of Russia
chief overland routes to the East
route of Khabarovsk to the Amur

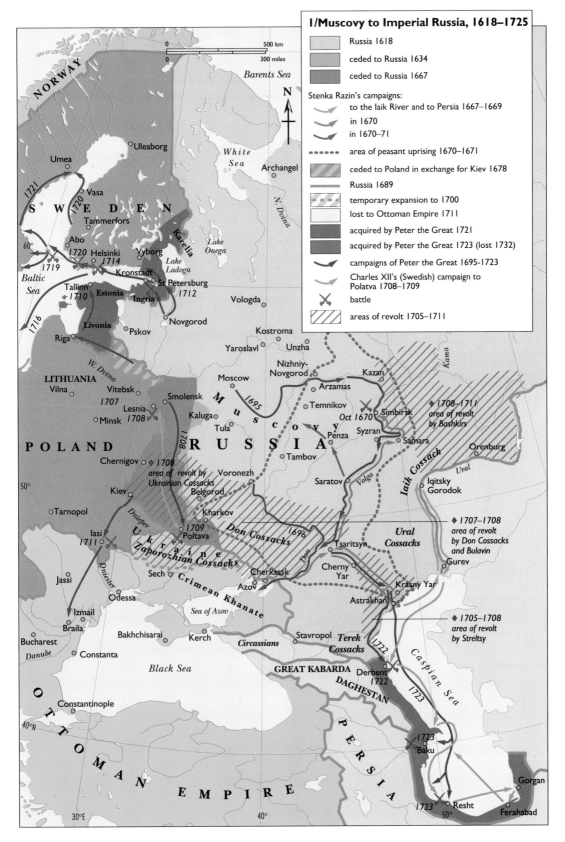

I/Muscovy to Imperial Russia, 1618–1725

Russia 1618
ceded to Russia 1634
ceded to Russia 1667

Stenka Razin's campaigns:
to the Iaik River and to Persia 1667–1669
in 1670
in 1670–71
area of peasant uprising 1670–1671
ceded to Poland in exchange for Kiev 1678
Russia 1689
temporary expansion to 1700
lost to Ottoman Empire 1711
acquired by Peter the Great 1721
acquired by Peter the Great 1723 (lost 1732)
campaigns of Peter the Great 1695–1723
Charles XII's (Swedish) campaign to Polatva 1708–1709
battle
areas of revolt 1705–1711

Russia to Catherine the Great

During a period of relative internal stability Russia, under the empresses, confirmed its position as a European power to be reckoned with.

"Pugachev stood in a long, raw sheepskin coat, almost struck dumb and beside himself. … He did not resemble so much a bestial and cruel brigand as some small victualler or shabby crook."
Andrey Bolotov, from his *Memoirs*

The period between 1725 and 1796 was dominated by the empresses Elizabeth I (reigned 1741–61), and Catherine II (the Great, reigned 1762–96). But it was in the reign of Anne (1730–40) that post-Petrine expansion began: in the war of 1736–39, Turkey was defeated and the recovery of Azov was complemented by the annexation of part of the southern steppes. Despite this success, the Crimean Tatars continued to harry Ukraine until the middle of the century.

Catherine's reign saw two further Turkish defeats and the Russian seizure of all the southern steppes to the Black Sea, and as far west as the River Dniester. The Crimea was captured and annexed in 1783. The newly-acquired territory was renamed New Russia and placed under the administration of Catherine's powerful favourite, Prince Potemkin, who ensured that settlement of these territories commenced immediately.

Expansion was not limited to the south; already a Russian protectorate in all but name, Poland was partitioned on three separate occasions: partially in 1772 and 1793, and finally in 1795. As a result Russia, for the first time, shared common frontiers with Austria and Prussia. These moves brought all of Belorussia, non-Slav Courland and Lithuania under Russian sway, as well as introducing a million Jews to the empire.

Although Russia enjoyed relative internal stability during most of the 18th century, Catherine's security was threatened by the Cossack rising led by Yemelyan Pugachev who claimed to be Peter III—the husband murdered on Catherine's orders in 1762. Originally on the River Yaik (now the Ural) the rising gathered momentum as it moved onward towards the Volga. Pugachev captured Kazan and during the winter of 1773–74 even seemed to threaten Moscow. Regular troops finally captured the rebel leader and his execution in Moscow secured Catherine's position.

Equestrian portrait of Catherine the Great by Vigilius Erichsen (1762). Though anxious to portray herself as an enlightened autocrat corresponding with such luminaries as Voltaire and Diderot, Catherine was also responsible for measures which would enable a landowner to send a serf to hard labour in Siberia and which effectively reduced serfs to actual slavery.

2/Popular Rebellion

- direction of peasants' flight
- original Cossack settlement by 1500
- route of Pugachev's campaign
- area of revolt 1773–1774
- town seized by Pugachev
- battle
- area under Cossack control by 1800

1/Russian Expansion in Europe, 1722–1796

- European (and West Siberian) Russia, 1725
- Bashkir rising, 1735–1740
- Russian boundary in Europe, 1796
- boundary of Russian province, 1750
- Persian territory held by Russia, 1723–1735
- annexations 1725–1762
- territory annexed to Russia by partitions of Poland
- annexations 1774–1786

Russia in Asia

Throughout the 18th century Russia continued to add to its possessions in Asia, strengthening its hold with the establishment of naval and military bases.

By 1735, nearly all of Russia's earlier annexations along the Caspian had been restored to Persia: all lands south of the Kura River in 1732, and the boundary, moved north to the region around Petrovsk in 1735. Yet elsewhere frontiers continued to advance. Omsk was founded on the middle Irtysh River in 1716 and Semipalatinsk much further upstream some two years later. Kamchatka was conquered following its discovery (or rediscovery) in 1697. Native resistance had been checked by 1732; in 1740 the naval base of Petropavlovsk was founded on the southeastern coast, roughly at the same time as the Kuril Islands were explored (from *c.*1711) and claimed by Cossacks for Russia.

Between the death of Peter the Great (1725) and the accession of Catherine the Great (1762), few changes occurred to the empire's map. In 1731, the Kazakhs of the Lesser (Younger) Horde nominally accepted Russian sovereignty, extending Russian influence almost to the shores of the Aral Sea. In about 1740 such sovereignty was also accepted by part of the Middle Horde. Firm Russian control over these nomadic Kazakhs, however, had to wait until the 19th century.

North of Kazakh territory, Russia achieved further permanent gains with the founding of Orsk (*c.* 1735) and Orenburg (*c.* 1743) on the Yaik River (renamed Ural, 1775), with the fort at Orenburg serving as a base for subsequent expeditions deep into Central Asia. Another fort—Iletskaya Zashita—was established in 1743, and the town of Petropavlovsk on the Ishim River in 1752.

It is also possible that Alaska may have been sighted first by Russians in 1732, the year when Fedorov and Gvozdev are said to have reached Cape Prince of Wales on the eastern shore of the Bering Strait. Other expeditions followed: Bering reached North America in 1741; the first of many merchant voyages there (for furs) was undertaken two years later. Exploration of the Aleutian Islands (discovered in 1741) also occurred at this time. More ambitious Russian plans for the exploitation of North America were to come the following century.

A map of Siberia showing Tibetan tribes, drawn for Vitus Bering in 1729. Danish by birth, Bering lent his name to the straits which separate Siberia from Alaska.

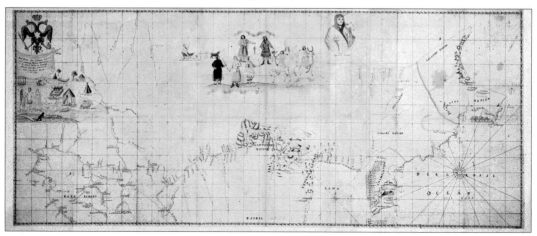

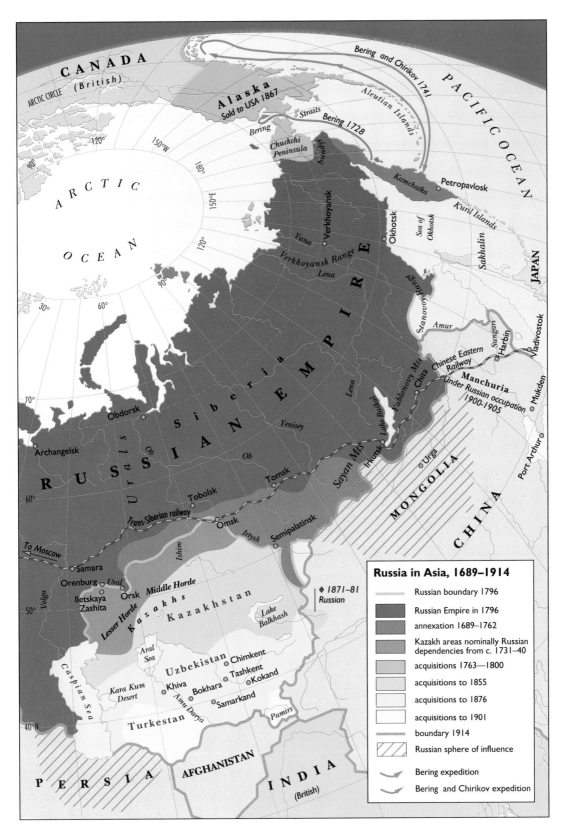

CANADA
(British)

ARCTIC CIRCLE

120°

90°

30°

60°

A R C T I C

O C E A N

Alaska
Sold to USA 1867

Bering
Straits

Bering 1728

Bering and Chirikov 1741

Aleutian Islands

Chuckchi
Peninsula

Anadyr

Kamchatka

PACIFIC OCEAN

Petropavlosk

Kuril Islands

Sea of
Okhotsk

Okhotsk

Sakhalin

JAPAN

150°W

180°

150°E

120°

90°

Yana

Verkhoyansk

Verkhoyansk Range

Lena

Stanovoy Range

Amur

Sungari

Harbin

Vladivostok

Chinese Eastern
Railway

Manchuria
Under Russian occupation
1900–1905

Mukden

Port Arthur

70°

Obdorsk

Lena

Yablonovy Mts

Chita

S i b e r i a n   E M P I R E

R U S S I A N

Urals

Ob

Yenisey

Lake Baikal

Irkutsk

Urga

MONGOLIA

CHINA

Archangelsk

60°

Ob

Tomsk

Sayan Mts

Tobolsk

Trans-Siberian railway

Omsk

Irtysh

Semipalatinsk

Ishim

To Moscow

Samara

Orenburg

Ural

Orsk   Middle Horde

Iletskaya
Zashita

Lesser Horde

Volga

♦ 1871–81
Russian

K a z a k h s

Kazakhstan

Lake
Balkhash

50°

Aral
Sea

Kara Kum
Desert

Caspian Sea

Uzbekistan

Khiva

Amu Darya

Chimkent

Tashkent

Bokhara

Samarkand

Kokand

Pamirs

Turkestan

40°N

P E R S I A

AFGHANISTAN

INDIA
(British)

Russia in Asia, 1689–1914

	Russian boundary 1796
	Russian Empire in 1796
	annexation 1689–1762
	Kazakh areas nominally Russian dependencies from c. 1731–40
	acquisitions 1763—1800
	acquisitions to 1855
	acquisitions to 1876
	acquisitions to 1901
	boundary 1914
	Russian sphere of influence
	Bering expedition
	Bering and Chirikov expedition

Provinces, Population and Economy

*By the end of Peter the Great's reign Russian industry had been
dramatically developed but both industry and agriculture
remained tied to the medieval system of serfdom.*

By the time of Peter the Great, serfdom had become well established, with
service obligations generally being discharged in one of two ways: labour
service or *barshchina*, involving an obligation to work so many days a week
on the landlords fields; and *obrok* which was either a fixed payment in kind
or in cash (or a mixture of the two). *Barshchina* was more common in the
south where the area farmed by the serf was usually only sufficient for subsistence purposes and the serf was highly dependent on the serfowner. In
the more northerly regions *obrok* predominated since the poorer quality
land meant that agriculture was usually insufficient for earning an income.
It was here that peasants engaged in a variety of trades or found employ-
ment off the farm. Some enterprising serfs—with obliging serfowners who
saw the gain for themselves—
eventually became serf entrepreneurs, fighting against tough
odds to obtain the serf's freedom
(at considerable expense to
themselves). Even under the constraints of serfdom, Russia was
not wholly without native private
enterprise.

Other peasants had managed to
avoid serfdom by a variety of
means. In 1724, Peter termed
them "state peasants", subject to

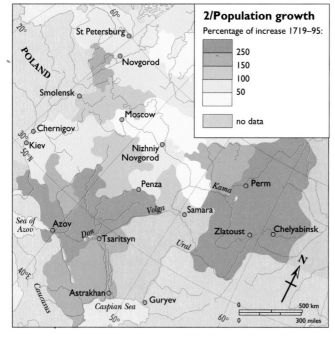

2/Population growth

Percentage of increase 1719–95:

- 250
- 150
- 100
- 50

- no data

3/Serfdom in 1678 and 1719

privately held serfs as a percentage
of the peasant population:

	1678	1719
	> 60	> 60
	44.3	65.7
	22.4	20.7
	0	3.4

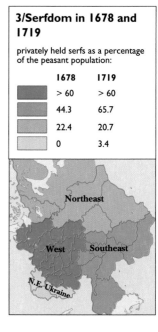

An 18th-century engraving reveals the conditions of peasant life.

state taxes, conscription, etc. As a consequence of the state's tendency to transfer more and more people into this category, it grew faster than that of proprietorial serfs. In 1724 there were *c.* 1.04 million male state peasants, almost 20 per cent of the empire's total male population. In the provinces of Vyatka, Kazan, Poltava, Tauride, Olonets, Voronezh and Archangel between 60 and 80 per cent of the population were state peasants.

By this stage Russian agriculture was already becoming regionally specialized, producing furs, timber, flax, rye, wheat, oats and barley, sugar beet, cotton, cattle, sheep, horses, fruit and wine.

Industry had also been spurred on under Peter, though the chief reason for industrial development was his intention to increase both the internal and external power of the state. Old industries were expanded, new ones created, and a substantial manufacturing sector developed while foreign commerce and trade grew. The industrial base was geared towards his military needs: cannon foundries, armaments factories and woollen cloth factories (supplying uniforms, sailcloth, rope etc.). Although the state built factories, Peter also drafted in individuals to set them up and manage them. Many foreign specialists were recruited.

To recruit labour for his new factories, Peter resorted to a variety of means. In particular, in 1721 he permitted merchants to purchase serfs for industrial labour, allowing them to purchase whole villages whose serfs became the property of the industrial enterprise, remaining with it even if the ownership changed. These so-called "possessional factories" extended serfdom, establishing a new industrial form. Thus, somewhat ironically, Peter's modern development simultaneously reinforced characteristics of economic and social backwardness.

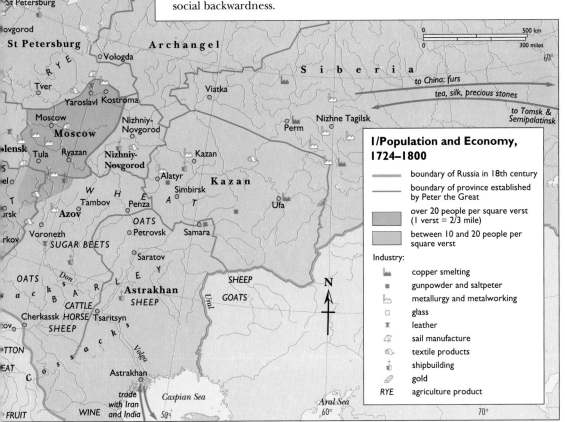

I/Population and Economy, 1724–1800

boundary of Russia in 18th century

boundary of province established by Peter the Great

over 20 people per square verst (1 verst = 2/3 mile)

between 10 and 20 people per square verst

Industry:

copper smelting

gunpowder and saltpeter

metallurgy and metalworking

glass

leather

sail manufacture

textile products

shipbuilding

gold

RYE   agriculture product

Peter the Great and St. Petersburg

In 1703 Peter the Great founded a city which was to become the most potent symbol of his westernizing policies.

"Petersburg, a city built rather against Sweden than for Russia, ought to be nothing more than a seaport, a Russian Danzig."
Marquis de Custine, *Russia,* 1754

Above: *Peter the Great by an unknown artist. Peter's loathing of Moscow and his desire for a new capital had their origin in the rising of the streltsy two weeks after the accession of the young tsar. For three days the old capital was rocked by the revolt of the soldier-traders while Peter's uncles and other connections on his mother's side were butchered. This event, linked to Peter's ambition to symbolize a new beginning for his westernized realm, gave birth to St. Petersburg.*

Right: *St. Petersburg from the River Neva. Built upon land captured from the Swedes, St. Petersburg became a powerful expression of Peter's military prowess and of his desire for naval supremacy. One of the first buildings to be completed was the Admiralty.*

To symbolize his confidence in the permanence of his victory over the Swedes and to exorcise his hatred of Moscow, in May 1703 Peter the Great founded a new city on the Baltic coast. Despite its bleak and inhospitable location, the single-minded Tsar determined that his new city, St. Petersburg, should become the capital of Russia and the showpiece of his westernizing policies. Construction began immediately with Peter paying little heed to the immense costs in both monetary and human terms as thousands of serfs were compulsorily conscripted for the fulfilment of the Tsar's dream. Working in freezing swamps in appalling conditions, many thousands of serfs died—providing another example of the Russian autocratic contempt for the lives of the peasantry.

Abandoning traditional Muscovite styles of architecture, St. Petersburg became a byword for classical regularity. The street plan, following a strict grid pattern, was indicative of Peter's ambition to emulate the western European cities which he had admired during his "incognito" tour of 1697–98. In 1712, St. Petersburg became Russia's new capital and, anxious to ensure a consistent rapidity of growth, in 1714 Peter banned the use of stone in buildings elsewhere. Soon the court and noble families were obliged to adopt the "Venice of the North" as their new home, and by the 1780s the city boasted a population of 200,000. By exercising his iron will, Peter had succeeded in transforming a dismal swamp into a potent symbol of both Russia's civic pride and military and naval ambition.

*The street plan of
St. Petersburg in the early
18th century (below). Peter's
desire to emulate the West is
revealed in the orderly street
grid and in the classical
facades of the public buildings
(left), a far cry from the
ancient onion-shaped domes
of the Moscow Kremlin.*

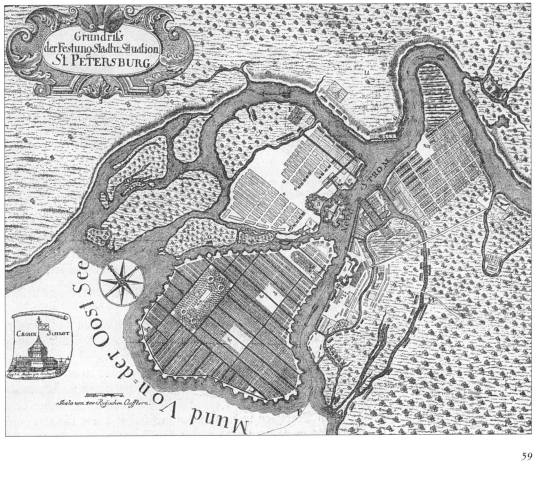

IV: Russia From Early 19th Century

Reaching a peak of power and prestige in Europe with Napoleon's defeat, the Russian tsars allowed themselves to be outstripped by the western powers due to an inability to modernize.

"... having prepared a strong army, Napoleon moved with all his forces and great quantities of arms onto our lands. Murders, fire, and devastation followed in his wake."
Manifesto of Alexander I, 3 November 1812

Russia entered the Napoleonic Wars against France as a member of the Coalition which comprised Britain, Austria, Naples, Portugal and Turkey. But Tsar Paul, disillusioned with Britain and Austria's failure to support Russia in the Netherlands, and believing that Napoleon would guarantee stability in Europe, abandoned the Coalition and joined France.

The war was to dominate the reign of Paul's successor, Alexander I. However, the disastrous battles of Austerlitz and Friedland—and Russia's heavy losses—led to Russia's withdrawal from the next phase of the war. In July 1807, Alexander concluded peace at a secret meeting with Napoleon at Tilsit. Now committed to supporting France, Russia was otherwise left free to take on other enemies. In 1808-1809, Alexander fought and defeated Sweden, the ensuing peace establishing Finland as an autonomous Grand Duchy of Russia. Declaring war on Turkey and Persia, Alexander won Bessarabia from the former, and Baku from the latter.

While France and Russia had benefited from the five-year break in hostilities, tension between them remained high. Peace between the two nations ended in 1812 when Napoleon's army crossed the River Beresina onto Russian territory and marched towards Moscow. Outside the city the battle of Borodino was fought with huge loss of life on both sides. The Russian commander, Kutusov, abandoned the city to Napoleon. The subsequent fire, which reduced Moscow to a ruin, left Napoleon no choice but to withdraw. On 19 October, in bitterly cold weather, the French army began its terrible retreat back to France. The Russians pursued Napoleon's depleted forces and, in 1814, occupied Paris.

Alexander was a prime mover in the Congress System of political alliances which was intended to maintain stability in Europe. By refusing to allow the dismemberment of a defeated France, setting up a Polish state under Russian control and upholding the supremacy of the autocratic states, Russia attempted under Alexander to dictate the future of Europe.

At home, Alexander—with the intention of creating a permanent, self-sufficient military class—established military colonies in a number of provinces. This involved entire peasant villages turning into army camps, some of which revolted in 1831 when there was a cholera outbreak. Alexander's preoccupation with international politics diverted his attention from reform. While a decentralising constitution was drafted by Count Novosiltsov, and a liberal constitutional system devised by Count Speransky, in practice very little was done.

Suddenly, at the age of 48, Alexander died. Confusion followed, not least as to who should succeed him—his brother Constantine or his youngest brother Nicholas. The announcement of the latter's accession in December 1826 was met with plans for a *coup d'état*. Those involved, mostly young army officers—aptly named Decembrists—assembled some 3,000 troops in St. Petersburg, and called for a constitution and Constantine instead of Nicholas as tsar. Forewarned, troops loyal to Nicholas fired into the crowd, killing 50 and arresting the leaders. Apart from the desire for a constitution and an end to serfdom, there was little consensus as to objectives, though the Decembrists were to become an important symbol to later generations of radical intelligentsia. As an opposition movement among Russia's upper classes, it was perceived as posing a dangerous threat to autocracy.

Later in Nicholas's reign, Mikhail Speransky codified Russian law, state finances were brought under control and economic growth came with an upsurge in grain exports. Nicholas preferred to rule through special committees. Political police, print censorship and tight restrictions on foreign travel were all features of his reign. Gendarme districts were established in 1837, as were special gendarme detachments at strategic towns, fortresses and ports. Conditions improved for state peasants as they were given titles to their land and effectively freed. By the mid-19th century over half of all peasants in the empire were not serfs (the property of private landowners) but "state peasants" who, although not free men, had more control over their lives than serfs. Serfdom itself, however, remained untouched. Pessimistic about their future, landlords transferred over 0.5 million serfs into domestic service during the 1840s and 50s in the hope of minimizing any distribution of their estates.

The Decembrist insurrection in Senate Square, St. Petersburg, 14 December 1825. The December uprising was of considerable importance as a symbol of the dissatisfaction of the Empire's educated classes. Their aim was not merely to unseat an individual ruler but to change the entire system of government in Russia, replacing the autocracy with a republic, or, at the minimum, a constitutional monarchy.

Nicholas's reign also began stormily in foreign matters, with two wars against Turkey, the first successful, the second unsuccessful. Newly established good relations with Turkey saw Russia defend Constantinople against the Egyptian rebel, Ibrahim Pasha; in 1848–49 Russia helped the Austrian Empire to crush the Hungarian national uprising led by Kossuth. The French revolution of 1848, and the subsequent accession of Napoleon III, made a clash with France more likely; the international power struggle that had developed between the major European states with the decline of the Ottoman Empire eventually resulted in the Crimean War (1853–56), which ended in a bitter defeat for Russia.

Nicholas died before the end of the conflict. Although his successor, Alexander II, initially desired to continue the war, new disasters made it

prudent to pursue peace at any price. In the Treaty of Paris (1856) Russia ceded Bessarabia to Moldavia, lost areas of the Balkan frontier, its naval rights in the Black Sea, the right to a sea passage into the Mediterranean and its protectorship over Orthodox Christians in the Ottoman lands. Not only had Russia's influence in Europe considerably diminished, but the war resulted in severe political, social and economic strains which were met by public resentment, especially among the intelligentsia. When he proclaimed an end to the war, Alexander hinted at the abolition of serfdom.

While serf unrest punctuated the early decades of the 19th century, this was probably exaggerated by earlier Soviet historians—the growing unrest among peasants during the Crimean War being directed not against the war and the government, but against serfdom and serfowners. The war led to rumours that new types of recruitment and civil guard duties would allow further freedom. (Lengthy military service was one way for serfs to gain their freedom, though few survived its harshness and brutality.) Rumours spread among peasants of a "promised land" somewhere in Russia, where serfdom did not exist, and this led to mass migration (eventually halted by force) in search of this utopia. Another rumour spread rapidly declaring that the new tsar had proclaimed an end to serfdom, but that the news was being suppressed by serfowners. This resulted in popular uprisings against serfowners.

The Crimean War had revealed the relative backwardness of the Russian economy when contrasted with the economies of western Europe, and highlighted the connection between economic progress, military advance and international political power. The long maintenance of serfdom retarded agricultural production, restricted freedom of movement and prevented large scale mobilization of peasants into an industrial labour force, thus slowing the development of industry. Linked to this were restrictions on civil and economic rights; the preservation of the privileged; the position of estate owners; continued autocratic government; and an administration largely based on the landowning class whose ideas pervaded official circles and most of the bureaucracy.

Blame for this backwardness has also been placed on the 250 years of Tatar rule, which resulted in a desire for military security and a strong state, as well as a strongly interventionist state in the economic sphere. The state, however, played a dual role: it both stimulated and impaired the development of science and technology; it tended to sponsor non-competitive enterprises through the creation of primitive monopolies and through a protectionist policy of excessive tariffs. Yet harsh methods of labour exploitation were unsuited to industry. Nepotism and corruption were now transferred to economic policy; bureaucratization retarded the growth of a commercial middle class; the requirements of industrial development were given second priority to war needs, the fulfilment of which were demanded by the old school military elite. This tradition determined the character of Russian industrialization after 1860.

The Kornilov bastion at Malakhov Hill during the Russian defence of Sevastopol, 1854-55. The fall of the fortress-city in September 1855 signalled the defeat of Russia and Alexander II quickly opened peace negotiations. The terms of the peace were finally ratified at the Congress of Paris in 1856.

Other factors also explain Russian backwardness: poor climate, the disadvantageous location for participation in international trade, a lack of ports, the dispersal of resources, and the country's immense size, all worked to limit unification of the minerals and fuels necessary for the development of heavy industry. The construction of railways—vital for reducing these problems—began in the 1840s with the St. Petersburg to Moscow line, but was not expanded until the 1870s and 80s. While it is usually accepted that Russia was a very rich country as far as natural resources were concerned, in reality European Russia was less endowed with such resources than most of the leading western European countries. For example, compared to the relatively favourable location of iron ore deposits in Britain, Germany and France—all countries with fewer transport difficulties—in Russia it was necessary to cross the Ural Mountains to find sufficient supplies of iron ore.

Climatic conditions, too, presented great difficulties—when compared with western Europe—for the introduction of advanced crop rotations, the breeding of livestock and the full utilization of labour in agricultural occupations. Given the same level of agricultural technique, the peasant of central and northern European Russia had to overcome more natural handicaps than his counterpart in western Europe. It is thus not so much a case of serfdom producing backwardness but backwardness reinforcing serfdom, already a symptom of backwardness.

The development of the Russian nation state was associated as much with tradition as with progress, and in state circles traditionalist views prevailed. Government policy towards industry was piecemeal, while railway expansion was deliberately delayed because of the fear that large movements of people would undermine social stability. However, ideas of economic progress were beginning to infiltrate the ruling élite during the early 19th century, especially among those high-ranking officials particularly concerned with economic policy.

The basis of education for a more modern society had already been laid in the reign of Alexander I. Technical research was conducted in the universities and the Academy of Sciences; engineering schools and technological institutes were founded; a public education system established. Russian scientists were part of the European scientific community, but Russia remained dependent on foreign engineers and equipment for the limited application of technology to industry that occurred before the 1860s. Some private industrial enterprises did make an appearance, bringing in investment capital and technology, and although relatively small in scale compared to the later 19th century, indicate the potential for autonomous growth. Similarly, while business, financial and commercial institutions were generally poorly developed, private insurance was making some headway, especially during the reign of Nicholas I. The first insurance company was established in 1827, and several large private insurance companies were operative by the 1850s. Not only did these generate investment capital, but provided some security against entrepreneurial risks, further encouraging private enterprise.

Expansion in the first half of the 19th century also occurred in trade, both at home and abroad. Although traditional fairs still predominated at home, more permanent urban markets were becoming established, and the home market was increasing for textiles and other goods. The European market was simultaneously increasing for Russian wheat. Many of the later large industrial enterprises had their origins in 1820–60 and these were highly concentrated, mechanized and utilized modern technology. Trade also benefited from extensions to the infrastructure which occurred in this period: the major canal systems of European Russia (having begun with Peter the Great) being built from the reign of Paul to that of Alexander II. These combined enormous feats of engineering with large amounts of labour and capital, much of it from abroad.

By 1860 Russian society was still largely traditional and agrarian in character, ripe for political, social and economic reforms. Alexander II came to power at a crucial moment in the mid-19th century. The Crimean War had revealed economic weaknesses, but also the lack of a trained army reserve that could be used to suppress peasant unrest during the war. The abolition of serfdom had made necessary a new system of military recruitment—more humanitarian and with a reduced period of service. Yet the delay in its implementation (until 1874) reveals that it was not just military problems that motivated Alexander to abolish serfdom. All reforms were interlinked. His now famous phrase that it is better to reform "from above" rather than wait for revolution "from below" is often seen as indicating his fear of the peasants, though he was also strongly influenced by social and humanitarian concerns.

The emancipation of 1861 was the central focus of the reforms (though even after its formal abolition, many features of a serf economy were preserved right up to the early 20th century). Peasants were granted personal freedom but farmed less land on average than before 1861; they had to pay for that land through annual redemption payments over 49 years (at a cost estimated at well in excess of the then market price); they were denied access to land which they had been able to use under serfdom. For both fiscal and administrative reasons, peasants were also tied to the village commune. It was clearly easier to administer many millions of peasants through communes than individually, especially when communal "mutual responsibility" ensured that all would pay their redemption dues and other taxes. Forcing peasants to pay for the land also spurred transport improvements—railways in particular—since to pay their redemption dues peasants needed to obtain cash, and for that they needed to sell their products.

Closely following the emancipation were the local government, legal and educational reforms. Liberation from the administration and jurisdiction of local landlords required a new system of local government. In January 1864, each rural district could elect representatives to a local council, or *zemstvo,* and assume responsibility for the maintenance of roads, bridges, poor relief, primary education and public health. To finance these, the *zemstvos* could levy taxes.

The *zemstvos* constituted a genuine form of local self-government. More enterprising ones also established hospitals, encouraged improvements in agriculture and commerce, engaged teachers, doctors, vets and agronomists. Many of these were liberal or radical in outlook, and this influenced their political orientation. The *zemstvos* often became centres of agitation for further reform by those who felt stifled by the centralized bureaucracy. With representatives at both district and provincial level, the next logical step was a national *zemstvo,* or Russian parliament, representing every district and class in the country. While not wishing to overthrow the existing government, they did want a greater share in political power and influence. In this context, they were perceived as posing a threat to autocracy and the bureaucracy, and made the latter extremely wary of them. Thus *zemstvos* were faced with official obstruction—such as reducing their powers of taxation—which meant they were constantly in financial difficulty. This was followed in June 1870 by re-organisation of the municipal administration; towns were granted a certain degree of self-government, representation being based on taxes and wealth. Yet, like *zemstvos,* while providing valuable experience in self-government and increased public services, the municipal adminstrations faced obstruction because of the perceived threat to government and bureaucracy.

The year of 1864 also saw important reforms affecting the legal system. Petty offences were no longer handled by the landlords but by JPs elected by *zemstvos.* More important cases were placed before higher courts where judges were appointed by the Crown. The legal code was revised in accordance with principles of western jurisprudence while court proceedings were to be conducted in accordance with a strictly regulated system of prosecution and defence. Trial by jury was introduced for criminal cases, and lawyers emerged as a profession, two of the best known products being Alexander Kerensky (later head of the Provisional Government in 1917) and Lenin. Yet peasants—the vast majority of the population—retained their own system of rural justice, centred in *volost* courts, based on customary law and procedures. Local government and legal reforms were underpinned by educational reforms producing the specialists required. *Zemstvos* helped expand primary education, while technical institutes and universities were granted autonomy and were influenced by western ideas.

A major by-product of the reforms was the creation of an intelligentsia which was left searching for an identity. Some found their vocation in the new local government organizations and worked to improve the condition of the people. Others took advantage of the opportunities offered by higher education to appease their consciences and purge themselves of the guilt they harboured for having a relatively privileged background. Others became radicals, and these ultimately turned to unrest in the 1870s—some "going to the people" in the villages only to be subsequently disillusioned and turning to terrorism (Populists). Others turned to Marxism.

By the 1870s the government was uncertain how to deal with such unrest. Loris-Melikov, the Minister of the Interior, appreciated that more severe police methods were inadequate, and so relaxed government censorship, believing there was a need for a determined effort to win over moderate advocates of social and constitutional reform (the government's natural allies against revolutionaries because they now had a "stake in the system"). The supreme irony was that moves were afoot to create a new legal channel for expressing opinions (in reality, another concession, a desperate move to preserve autocracy at any price) when, on 13 March 1881, Alexander II was assassinated. After his father's violent death it is not surprising that Alexander III (1881–94) concluded that reform of any kind was a mistake. The late 19th century was a period of "reactionary policies", "counter reform" and Russification.

Alexander II attending the signing of the Treaty of San Stefano, March 1878.

Russia in Europe after 1798

U*ncertainty as to Tsar Paul's allegiance led first to the French occupation of Moscow, and then to the Russian capture of Paris.*

"So I'm finally in Moscow, in the ancient palace of the tsars, in the Kremlin!"
Napoleon, 15 September 1812

Although Catherine's successor, her son Paul I, began his reign desiring peace, in 1798 Russia joined Britain, Austria, Naples, Portugal and Turkey in the Second Coalition, and entered the war against France. In subsequent campaigns, a Russian fleet seized the Ionian Islands from France and created a Russian-controlled republic under Turkey's protection.

Russia's most successful operation during this stage of the war occurred in Italy during 1798–99, when a force of 18,000 Russians and 44,000 Austrians, led by Alexander Suvorov, finally drove the French out of Italy. Ordered not to invade France, Suvorov instead conducted a brilliant retreat through the Swiss Alps and into southern Germany.

In 1800, disillusioned by the failure of Britain and Austria to support Russia in the Netherlands, Paul abandoned the Coalition, and joined France. But the disastrous battles of Austerlitz and Friedland brought home to Paul's successor, Alexander I, the extent of potential losses and Russia withdrew from the war. Alexander concluded a peace with Napoleon at Tilsit in July 1807, which committed Russia to support France. Thereafter, however, tensions mounted between the two nations, culminating in the French invasion of Russia in June, 1812.

After defeating the Russians at Borodino, Napoleon took the fatal decision to advance on Moscow. Although unable to hold the city, neither Alexander nor his commander, Kutusov, had any intention of surrendering. Napoleon's entry into Moscow had come too late: it was September, the troops were short of supplies and there was no popular uprising in support of the self-proclaimed liberator. The retreat from Moscow began in mid-October. Decimated by the bitter cold, disease, hunger and constant harassment, the French were driven from Russian soil. The Russian army pursued Napoleon and occupied Paris in 1814.

The year 1812 retains a special place in Russian history and was to have important political and psychological consequences. Believing in divine approval, Alexander set out to control Europe's destiny. He was a prime mover in the Holy Alliance, and the Congress System of political alliances which was intended to maintain the balance of power and keep peace in Europe.

GREAT BRITAIN

London

Li

Paris

Seine

Loire

FRANC

Pyrenees

Ebro

SPAIN

Barce

0°

The Battle of Borodino, painted by Rubo. Sixty miles west of Moscow Napoleon hurled his forces against the Russian defenders under Kutuzov. Both sides experienced devastating losses and battered each other to complete exhaustion.

Russia moves West

	European Russia 1796
	Russian naval campaign 1798–1799
	Suvorov's campaign into Italy and Switzerland 1799
	war with Sweden 1801–1809
	Russian campaign against Turks 1806–1812
	Napoleon's advance on Moscow 1812
	Napoleon's retreat from Moscow 1812

	Russian advance on Paris 1812–1814
	Advance of Allied Forces on Paris 1813–1814
✕	Battle
	Russian acquisitions by 1812
	Russian acquisitions by 1815
	Russian boundary in 1815
	European frontiers in 1815
	German Confederation
●	International conference centre
	national revolution denied Russian support

White Sea
65°

Uleaborg

Umea

Vasa

Finland
(to Russia 1809)

Lake Onega

Lake Ladoga

60°

NORWAY

SWEDEN

Abo

Helsinki

St Petersburg

Stockholm

Novgorod

Baltic Sea

Riga

Moscow
55°

Borodino

Vyazma

Maloyaroslavets

Smolensk

Tula

orth Sea

DENMARK

Tilsit

Nieman

Vilna   Studenka

Krasnoy

Mogilev

Minsk

R U S S I A

Bobrov

Friedland
1807

Eylau
1807

Bialystok
(to Russia
1807)

Vistula

Pavlosk
50°

HANOVER

PRUSSIA

Poland
(to Russia 1815)

Kiev

Ekaterinoslav

ix-la-
hapelle

1813

Leipzig

Breslau

SSIA   Small states   SAXONY

Prague

Lemberg

Tarnopol
(to Russia 1809
returned to Austria 1815)

BAVARIA

Carlsbad

Cracow

Troppau

Kamenets-Podolsk

Dnieper

ourg

Bohemia

Austerlitz
1805

Carpathians

Bessarabia
(to Russia 1812)

Dniester

Nikolaev

BADEN   BAVARIA

Danube

Vienna

Buda   Pest

Moldavia

45°N

SWITZERLAND

Austria

AUSTRIAN EMPIRE

Hungary

Alps

Sevastopol
35°E

DINIA   Milan

Laibach

Wallachia

Bucharest
1811

Izmail

Novi

Bosnia

Belgrade

Danube

Serbia

Nicopol

Varna

TUSCANY   PAPAL
STATES

Ancona

Bulgaria

Corsica

Rome

Cattaro

Constantinople

40°

Naples

Macedonia

Sardinia

Pindus

OTTOMAN EMPIRE

KINGDOM OF
THE TWO
SICILY

1799

Corfu

1798

Palermo   Messina

Ionian
Islands

Athens

Aegean
Sea

Dodecanese
to Alexandria

diterranean   Sea

5°   10°   15°   20°   25°   30°

European Russia 1801–1881

Autocratic repression led to a growth of revolutionary activity which culminated in the assassination of the liberator tsar.

"A tsar should be a good shepherd, ready to devote his life to his sheep. Alexander II was a ravening wolf and a terrible death overtook him."
Statement issued by the assassins of Tsar Alexander II, 13 March 1881

The period between 1801 and 1881 is one dominated by extremes of reaction and reform. At the beginning of the century any reformist moves ceased with Alexander I's belief in his divine destiny, and his absorption in high international politics and fringe religion. Although a decentralizing constitution was drafted little was done in practice about implementing it, or about serfdom. In the Baltic provinces, however, peasants were liberated but without land, which effectively impoverished them further. While Alexander's reign did exhibit some liberal leanings—such as numerous high schools and no fewer than five new universities were founded—it was also marked by maladministration and repression. One of the Tsar's most eccentric and unpopular innovations was the establishment of military colonies which resulted in entire peasant villages being turned into armed camps. All male adults under 45 had to wear military uniform and all children over seven were given military training as part of Alexander's plan for the creation of a permanent, self-sufficient military class.

The accession of Nicholas I (tsar 1825–55) was heralded by an attempted *coup d'état* led by young army officers and gentlemen intent upon constitutional government. The plot of the Decembrists, as they became known, was betrayed, and its leaders executed or exiled to Siberia. Though a failure, the

2 /Polish revolts 1831, 1861-63

— eastern frontier of Poland before 1772

— international frontiers 1830

▨ Congress Poland

▨ area of revolt outside Congress Poland

➘ Russian offensive February 1831

➘ Russian offensive April-Sept 1831

➘ Polish troop movements

• centre of the Polish revolt 1861-1863

➘ Prussian and Austrian troops aid Russia to suppress uprising

Poland ruled by the Russian Tsar 1815-1914

December plot provided a potent symbol of struggle to later generations of radical intelligentsia. Although Nicholas was intent upon crushing all revolutionary activity—as exemplified by his savage suppression of the Polish insurrection in 1831—and ran his empire in a military fashion, Russian laws were codified, order brought to state finances, economic growth experienced and the lot of state peasants improved during his reign.

The Crimean War (1853–56), which concluded Nicholas's reign, did much to convince his successor that wide-ranging reforms must be implemented if Russia was to retain its place as a European power. Alexander II (tsar 1856–1881) immediately began a relaxation of the controls imposed by his predecessors, though by crushing a second Polish revolt in 1863 Alexander revealed there was a limit to his reforming zeal. The most important single event in Alexander's reign was the emancipation of the serfs in 1861. Though an act of momentous significance, many of the Tsar's subjects remained dissatisfied with the new reforms, calling for more land for emancipated serfs and greater relaxation of the remaining laws. Increasingly active revolutionary groups plotted the Tsar's death; their plans were crowned with success on 1 March 1881 when Alexander II was assassinated.

I /European Russia, 1810–1831

- Russia 1825–1881
- military colonies established by Alexander I 1810–1825
- serfs liberated without land rights 1816
- peasant uprising 1826–1827
- peasant uprising provoked by the drought of 1847
- centre of Decembrist uprising
- areas of revolt in military colonies ravaged by cholera 1831

Economic Development to 1860

Failing to appreciate the need for economic development, Russia was rapidly outpaced by its western European rivals.

> *"I am convinced, ... that sooner or later we shall have to grant emancipation. ....it would be very much better if it were to come from above rather than from below."*
>
> Alexander II,
> March 1856

By the beginning of the 19th century Russia had developed to such a degree that it enjoyed the rank of one of Europe's greatest powers—and yet in many ways it lagged far behind its western European rivals.

The hundred years between the end of the first quarter of the 18th century, and that of the 19th, had witnessed a remarkable growth in the Empire's population. In 1724, at the end of Peter the Great's reign, the Russian population had numbered around 14 million; in 1835 this had risen more than fourfold to 60 million. In 1800 the vast majority of the tsar's subjects remained rural-dwellers

According to the 1858 census, there were some 10.7 million male serfs out of a male population of 24 million. Forty-six per cent of all serfs were concentrated in just 18 provinces of three regions: the Central Industrial, the Central Black-Earth and the Lakes. Serfs performed two basic types of obligation; *barshchina* or labour services, indicated that serf labour in agriculture was important, whereas in provinces where *obrok*, or dues in kind, cash, or a mixture of the two, were more popular, agriculture was not so important to landlord income. Growth of trade and industry enabled more serfs to earn a living from work in commerce and manufacturing. They were in a better position to pay cash *obrok* than when entirely dependent on agriculture.

Continued territorial expansion meant that by 1855 the Russian Empire covered 7.8 million square miles, but the system of transport and communications remained wholly inadequate. The two premier cities of St. Petersburg and Moscow were not linked by a metalled road until 1830, and

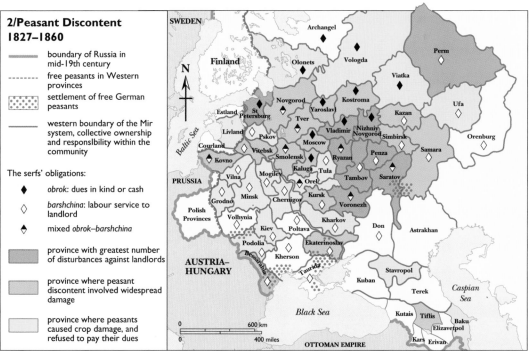

2/Peasant Discontent 1827–1860

— boundary of Russia in mid-19th century

--- free peasants in Western provinces

▦ settlement of free German peasants

— western boundary of the Mir system, collective ownership and responsibility within the community

The serfs' obligations:

◆ *obrok*: dues in kind or cash

◇ *barshchina*: labour service to landlord

◆ mixed *obrok–barshchina*

▨ province with greatest number of disturbances against landlords

▨ province where peasant discontent involved widespread damage

▢ province where peasants caused crop damage, and refused to pay their dues

I/The Russian economy by 1860

- Russia mid–19th century
- railway built by 1860
- railway under construction in 1860
- factory development before 1860
- town with large factory growth from 1860
- area of iron and steel production
- sugar factories

Industry expanding rapidly:
- leather
- wool
- linen
- tobacco
- oil
- coal
- copper
- gold

····· area where landserfs sold to factory owner, revolted and were suppressed by force

Serfs as a percent of the population:
- 50
- 30
- 10

it took another 21 years for them to enjoy the benefits of railway communication. Unfortunately, these developments did not set a trend, and the overall inadequacy of the system was demonstrated during the Crimean War (1853–56). The movement of commodities such as grain remained almost wholly dependent upon barges pulled by human teams, with the development of steamer traffic remaining insignificant until the 1850s.

Industrially, some progress was enjoyed, though this was seldom developed to its full potential. Opportunities for expansion were neglected, and in this sphere, as in so many others, Russia remained painfully backward in comparison with such powers as Britain, France and Prussia. In the 18th century, for example, Peter the Great founded the mining and iron industries in the Urals, enabling Russia to become, by 1800, the world's largest producer of pig iron. But rather than being encouraged, the industry was allowed to stagnate, indicating Russia's failure to appreciate the increasingly important bond between economic progress and political prestige on the world stage.

Russia in the Caucasus

A coherent policy of expansion in the Caucasus enabled Russia to defeat Persia and Turkey, and to capture substantial territories.

The old quarter of Tbilisi, the capital of Georgia, is overlooked by the ruins of the Arab castle on Mount Mtatsminda. Founded in AD 455 or 458, Tbilisi is one of the world's most ancient surviving cities. Captured successively by the Persians, Byzantines, Arabs, Mongols and Turks, the city accepted Russian domination in 1801.

Although Russia's boundaries stretched to the Terek River and Kabarda (north of the Caucasus) by the 16th century, the areas between the Black and Caspian Seas were not subjugated until the period between 1783 and 1878. Discontent among native peoples and pressure from oppressive Turkish and Persian neighbours facilitated Russia's early acquistion of the central mountain region, namely Georgia.

Moving south from Mozdok (founded in 1763) Russia established contact with Georgia via the Daryl Gorge on the Upper Terek River. Vladikavkaz was founded (1784) midway between Mozdok and the Daryl Pass, serving as the northern terminal of the Georgian military road leading to Tiflis (Tbilisi). In 1783, eastern Georgia voluntarily accepted Russian protection and in 1801, fearing Persian invasion, was formally incorporated into the Russian Empire. The annexation of most of the disunited western states—once part of ancient Georgia—followed swiftly, being completed in 1810.

Persia resisted such incursions into its sphere of influence, but lost the war with Russia that followed (1804–13). The Treaty of Gulistan gave Russia control of Georgia and ceded Dagestan and northern Azerbaijan.

In 1817, the Russian border was pushed south from middle Terek to the Sunzha River, where a new military line was established with a fortress (modern-day Grozny) founded in 1818. The subjugation of Dagestan was finally completed in 1859. The Treaty of Turkmanchai ended a war with Persia (1826–28) giving Russia the eastern Armenian Khanates of Yereran and Nakhichevan. By the Treaty of Adrianople in 1829, Turkey ceded Akhaltsikh province which was united with Georgia; it awarded to Russia the entire northeastern littoral of the Black Sea with the ports of Anapa, Sukhumi and Poti.

Final gains in Transcaucasia resulted from the Russo–Turkish war of 1877–78. The Treaty of San Stefano brought to Russia the districts of Batumi, Kars, Ardahan and Bayazid (Bayazet). The Congress of Berlin (1878) confirmed most of these acquisitions, although the Bayazid area was returned to Turkey, and Batumi was termed a free port under Russian jurisdiction. This settlement confirmed Russian borders in Transcaucasia until the outbreak of World War I.

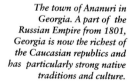

The town of Ananuri in Georgia. A part of the Russian Empire from 1801, Georgia is now the richest of the Caucasian republics and has particularly strong native traditions and culture.

46°N
Sea of Azov

Anapa
Novorossiisk
Ekaterinodar
Maikop
Labа
Kuban

R U S S I A N   E M P I R E

Volga

Stavropol
Astrakhan

44°

Circassians
Abkhazia
Kuban

Kislovodsk

Black
Sea

Sukhumi
Karachai
Kabarda
Mozdok
Terek

42°

Svanetla

Mingrelia
Ossetia
Vladikavkaz
Sunzha
Grozny

40°E
40°

Poti
Kutaisi
Imeritia
Daryl Gorge
Chechnya

Guria
Batumi
Trabizond

Gori
C a u c

Akhalsikh
Tbilisi
Tarki
Mekhtulinsk

Caspian Sea

Ardahan
Ardahan

Dagestan

G e o r g i a
Kura
Ilisuyski

Derbent

Alexandropol
Kars
Araks
Shuragei

OTTOMAN

EMPIRE

A r m e n i a
Yerevan
Lake Sevan
Gandzha
Elizavetpol

Sheki
Kusa

Yerevan

Bayazid

A z e r b a i j a n
Shirvan
Baku
Baku

Nakhichevan
Karabakh
Kura

42°
44°
Nakhichevan
46°
Araks

PERSIA

Talysh
Tabriz
Lenkoran

Astaria

48°
50°

Russian Expansion in the Caucasus from 1763

Russian Empire, 1763	annexed 1828
annexed 1783	annexed 1829
annexed 1786	annexed 1800–1830
annexed 1801	annexed 1830–1864
annexed 1813	annexed 1878 (ceded to Russia by Turkey)
dependent in 18th century, annexed first half 19th century	Russian boundary 1878–1914

Russia in Central Asia

As the Russian Empire swallowed increasingly large areas of Central Asia, tensions between Russia and Britain mounted.

An Uzbek artisan spins yarn for commercial use during the early years of the 20th century.

The period of 150 years between 1763 and 1914 was one of conflict and expansion as much of Central Asia was brought under Russian domination. During the 1740s, Russia established nominal control over the Lesser and Middle Kazakh Hordes and when the Elder Horde was eventually subjugated (*c.* 1847), the Uzbek rulers to the south of the Kazakh areas found themselves neighbours of the expansionist Russian Empire. Bitter conflict stretching over the next three decades resulted in Russia's domination of Kokand, Bokhara and Khiva, and the gradual establishment of advance bases for the planned conquest of the Uzbek territory. In 1853 Ak-Mechet (renamed Perovsk) in northwestern Kokand fell, followed by Chimkent, and in 1865 the capital at Tashkent. By the peace treaty of 1868, Kokand became a Russian protectorate (prior to its complete annexation in 1876).

In 1865 Bokhara's forces unwisely attacked the Russians in Kokand and the resulting war ended in 1868 with the loss of their capital, Bokhara. Peace was followed by the Russian annexation of Samarkand and adjacent areas and the remainder of the emirate of Bokhara became a Russian protectorate (until 1917). The Khiva Khanate, the least accessible of the Uzbek states, was the last to fall. The Russians annexed the right bank of Khiva's sector of the Amu-Darya River and the remainder of the Khanate, like neighbouring Bokhara, remained a protectorate until 1917.

Besides bringing Russia into conflict with the Khanates, the policy of expansionism in Asia raised fears in Britain over Russia's possible designs on India. With each annexation the likelihood of war increased. After the incorporation of the Uzbek areas into Turkestan, centred at Tashkent, Russia began the conquest of the isolated Teke-Turkmen settlements south of the Kara-Kum Desert .

Affluent-looking cotton merchants weigh their merchandise at a cotton-ginning mill in Tashkent.

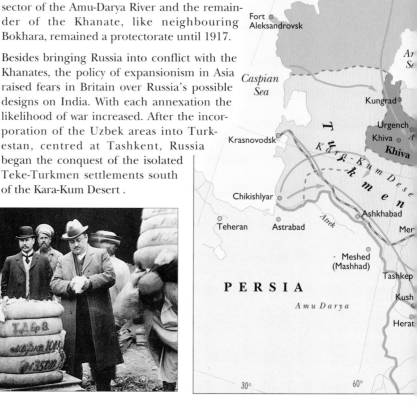

The last addition of contiguous territory was the rugged Pamirs sector annexed in 1895. This area constituted the Russian Empire's closest point to British India, the two being separated by Afghanistan—a buffer zone only ten miles wide in places. In an attempt to reduce mounting tensions, the Wakhan Mountain territory was ceded to Afghanistan by Britain and Russia in 1905. Such attempts at conciliation reached their zenith in 1907 with the Anglo-Russian Convention in which Russia, weakened by internal discord and defeat in the Far East, and Britain, increasingly aware of Germany as a potential enemy, sought to defuse the causes of antagonism.

Overall, this period is marked by a massive acquisition of territory which helped to ensure Russia's position as a colonial power, despite the disappearance of its North American overseas empire with the sale of Alaska to the United States of America in 1867.

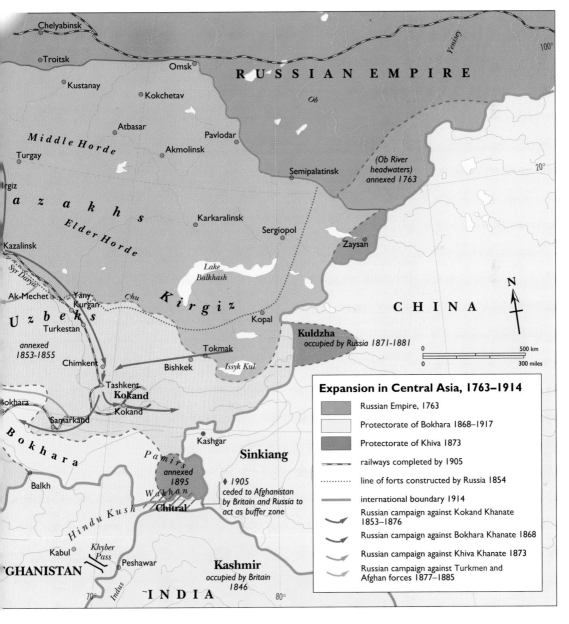

Expansion in Central Asia, 1763–1914

- Russian Empire, 1763
- Protectorate of Bokhara 1868–1917
- Protectorate of Khiva 1873
- railways completed by 1905
- line of forts constructed by Russia 1854
- international boundary 1914
- Russian campaign against Kokand Khanate 1853–1876
- Russian campaign against Bokhara Khanate 1868
- Russian campaign against Khiva Khanate 1873
- Russian campaign against Turkmen and Afghan forces 1877–1885

The Crimean War

Intent upon maintaining its influence within the Ottoman Empire, Russia embarked upon a war which was to end in defeat and humiliation.

"What is beginning is not a war; not a policy: it is the birth pains of a new world ... it is the decisive battle of the West and Russia."

F.I. Tyutchev, Russian lyric poet

The origins of the Crimean War lay in the power struggle between Russia, Austria, Britain and France for influence over the declining Ottoman Empire, Russia's primary concern being the neutralization of Turkey and the securing of a passage from the Black Sea into the Mediterranean. Britain and France, meanwhile, were fearful of Russian expansionist policies aimed at the Middle East.

By the 1850s suspicion between the interested powers had waxed to unprecedented heights becoming focussed on the question of who had jurisdiction of the Holy Places of Turkish-controlled Jerusalem. In 1852 the French government persuaded the Turkish authorities to cede to the Roman Catholic Church custody of a church in Bethlehem, despite the fact that the Greek Orthodox Church was the traditional custodian. Nicholas I, as patron of the Orthodox Church under Ottoman rule, demanded its return. When Turkey refused, he ordered the occupation of Moldavia and Wallachia. A Turkish declaration of war in 1853 was followed on 30 November by a naval engagement at Sinope on the Black Sea, in which a Turkish flotilla was destroyed. Russia's continued refusal to withdraw its forces resulted in Britain and France declaring war and sending an expeditionary force to the home port of the Russian Black Sea Fleet, Sevastopol, in southern Crimea.

Meanwhile, having made a defensive pact against Russia with Prussia, Austria moved troops into Moldavia and Wallachia, forcing a Russian withdrawal in 1854. Although the Anglo–French expedition to the Crimea was marked by incompetence, symbolized by the ill-fated charge of the British Light Brigade during the Battle of Balaclava, it still succeeded in defeating, on its home territory, the equally incompetent army of Tsar Nicholas. Russian forces were beaten at Balaclava and Inkerman and the capture of the Malakov and Redan strongholds brought to an end the year-long siege of Sevastopol.

The succession of Alexander II in 1855 was quickly followed by the opening of peace negotiations, which culminated in the Treaty of Paris (1856). With its influence in Europe clearly on the decline, Russia was forced to cede Bessarabia to Moldavia, and to agree to the neutralization of the area around the Black Sea. Areas on the Balkan frontier, naval rights in the Black Sea, right of sea passage into the Mediterranean and protectorship over Orthodox Christians in Ottoman lands were all lost to Russia.

Tsar Nicholas I was obsessed with enforcing a military-style discipline within the empire's bureaucracy. He died in the midst of a war which revealed the inadequacy of his army and the hopelessness of any attempt to compete with the economically-advanced states of western Europe.

Besides territorial losses, the Crimean War had a number of consequences within Russia. The disastrous performance of the Tsar's army resulted in a loss of prestige at home as well as abroad. Heavy wartime taxation, the burden of which fell particularly on the urban populace, caused severe economic disruption and widely-felt dissatisfaction. Popular demonstrations occurred in Moscow; when proclaiming the war's end, Alexander went so far as to hint at the abolition of serfdom. Above all, Russia had been made painfully aware of its economic backwardness in comparison with western European powers, and the inferred relationship between economic progress, military prowess and international political power.

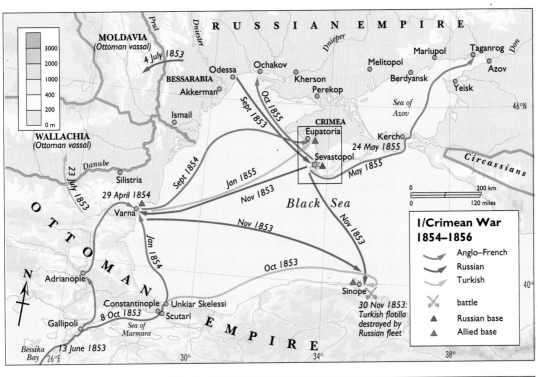

MOLDAVIA
(Ottoman vassal)

RUSSIAN EMPIRE

Dniester

Dnieper

4 July 1853

Odessa

Ochakov

Kherson

Perekop

Berdyansk

Melitopol

Mariupol

Taganrog

Azov

Yeisk

Don

Sea of
Azov

46°N

BESSARABIA

Akkerman

Ismail

Sept 1853

Oct 1855

CRIMEA

Eupatoria

Sevastopol

24 May 1855

Kerch

May 1855

Circassians

WALLACHIA
(Ottoman vassal)

Danube

Silistria

29 April 1854

Varna

Sept 1854

Jan 1855

Nov 1853

Black Sea

Nov 1853

Nov 1853

200 km

120 miles

O T T O M A N

Adrianople

Jan 1854

Nov 1853

Oct 1853

40°

N

Constantinople

8 Oct 1853

Unkiar Skelessi

Scutari

Gallipoli

Sea of
Marmara

E M P I R E

Sinope

30 Nov 1853:
Turkish flotilla
destroyed by
Russian fleet

Bessika
Bay

13 June 1853

26°E

30°

34°

38°

**1/Crimean War
1854–1856**

Anglo–French

Russian

Turkish

✗ battle

▲ Russian base

▲ Allied base

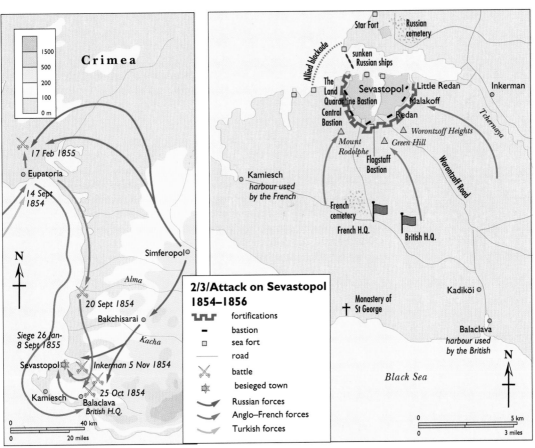

Crimea

1500
500
200
100
0 m

✗ 17 Feb 1855

Eupatoria

14 Sept
1854

Simferopol

N

Alma

20 Sept 1854

Bakchisarai

Kacha

Siege 26 Jan–
8 Sept 1855

Sevastopol

Inkerman 5 Nov 1854

Kamiesch

25 Oct 1854

Balaclava
British H.Q.

0   40 km

0   20 miles

**2/3/Attack on Sevastopol
1854–1856**

ᒐᒐ fortifications

— bastion

□ sea fort

road

✗ battle

✸ besieged town

Russian forces

Anglo–French forces

Turkish forces

Star Fort

Russian
cemetery

Allied blockade

sunken
Russian ships

The
Land
Quarantine Bastion

Central
Bastion

Sevastopol

Little Redan

Malakoff

Redan

Inkerman

Tchernaya

Worontzoff Heights

Green Hill

Mount
Rodolphe

Flagstaff
Bastion

Worontzoff Road

Kamiesch
harbour used
by the French

French
cemetery

French H.Q.

British H.Q.

Kadiköi

Monastery of
St George

Balaclava
harbour used
by the British

N

Black Sea

0   5 km

0   3 miles

V: From Reform to Revolution

Racked by the growing pains of industrialization and the increasingly urgent calls for political reform, Tsarist Russia blundered on into the jaws of World War I.

"One had the same feeling as before a thunderstorm in summer."
Nicholas II, 1905

The last two decades of the 19th century witnessed remarkable growth in the mining, metallurgical, textile and chemical industries in Russia, which far outstripped growth in other European states. By 1914, the Russian Empire had become the world's fifth largest industrial power, and the world's fourth greatest producer of pig-iron, steel, coal and cotton. Its railways had expanded greatly since the 1860s and the Baku oilfields were second only to those of Texas. Although Russia still had to import machine tools and electrical goods, Russian factories were nevertheless producing the industrial materials essential to make the Empire self-reliant. But the Empire still had to overcome a number of social, political and economic problems if it was to compete successfully with its industrial rivals in western Europe.

Its vast expanse of territory made Russia the largest state in the world. Despite the abundance of natural resources, size was not necessarily an advantage since the country suffered from a number of communication problems. Raw materials were widely dispersed and often far from the economic centres. Since the majority of navigable rivers flowed away from these centres, and many were frozen over and unusable for at least half the year, an increase in the country's rail network was far more pressing than it had been in other European states. The total length of track grew from about 1500 kms in 1860 to over 30,000 kms in 1890. Railway development also attracted state investment which, in turn, stimulated the economy. Two-thirds of the railway network was controlled by the state before the outbreak of World War I.

The Russian government's financial policies, and its orders for railways and armaments, were important to the country's economic development and stimulated industrial growth. Economic benefits could even be drawn from apparent setbacks. The destruction of the Baltic fleet at Tsushima in May 1905, although a disaster in terms of naval prestige, nevertheless was of great benefit to the ship-building, armaments and metallurgical industries and allowed for the construction of more up-to-date ships. The industrial economy was stimulated further in 1897 when the rouble was put on the gold standard. This attracted further investment from European entrepreneurs who had already been attracted by cheaper labour rates in Russia. In 1913, the net inflow of foreign investment was 578 million roubles, some 25 per cent of net domestic investment.

The attitudes of Alexander III were largely shaped by the assassination of his reforming father. The murderers had intended that their actions should so intimidate the new tsar that he would meet their demands for increased liberalization, but instead Alexander became reactionary, bitterly opposing any concession to the demands for further reform.

The process of industrialization in Russia reflected that of the major western European states at an earlier period. A gradual movement was developing towards the formation of an hereditary working class, where industrial workers and their families resided in urban centres, though by 1914 the majority of workers were still seasonal migrant labourers, moving between town and country. The conditions of Russian factory workers in general compared unfavourably to west European standards. They worked long hours for low wages in harsh factory conditions, where safety regulations were ignored, and often lived in overcrowded and insanitary, barrack-like buildings or urban slums. But these conditions were not unique to Russia.

Admittedly, reforms concerning conditions and hours in factories and mines had developed in Britain and Germany earlier in the 19th century, but even in Belgium in the 1880s it was common practice for women and children to work underground in the coal mines. Nevertheless, reforms were gradually being introduced in Russia, and by 1897 the working day had been restricted to eleven and a half hours. By the end of the century a rudimentary factory inspection system had been established. But Russia still had a long way to go to improve working conditions

Discontent came as a reaction to state intimidation. Russia was a highly authoritarian, centralized state and this was manifested through repressive internal security measures, strict censorship and a ban on trade unions and political parties. Russian workers at the turn of the century did not have the right to strike, and although some trade unions had been legalized, these had been infiltrated by police informers, (following the initiative of the St. Petersburg police chief, Sergei Zubatov), as a means of controlling any revolutionary tendencies.

Yet despite growing discontent, the 1905 revolution came as a shock to the socialist parties, and it seemed that the charismatic leadership of individuals such as Father Gapon had more effect upon events than the revolutionary ideas held by those belonging to a wide variety of Russian socialist groups. Although Lenin might have later called the events of 1905 "the great dress rehearsal" for 1917, the Social Democrats played no great part in it.

Lenin's time had not yet arrived; his activities for the time being were confined to building up an underground organization with a view to the future. The Bolsheviks had become a closed, disciplined and secretive élite of professional revolutionaries, nurtured in the spirit of Lenin's key pamphlet *What is to be done?* They were austere, resolute, suspicious and skilled in deceit, cut off from normal social life, constantly on the move, living with the danger of arrest and pursued by police spies. The Russian Social Democratic Party, and particularly the Bolshevik faction, was moving away from the ideas and policies of contemporary west European socialists, preferring conspiratorial techniques to mass, open party activities. No doubt

Below: Grand Duke Sergei, cousin of the Tsar and commander of the Moscow military region, meets his death at the hands of the "Battle Organisation" of the Social Revolutionary party in early 1905.

Right: Lenin and other young Marxist revolutionaries pose for the camera in St. Petersburg, 1897. The Marxist Social Democrats, as the group was known, held their earliest meeting in a log cabin near Minsk. Although aware of the party's existence, at first the Tsar's police did nothing, believing that the group concerned itself with economic theory rather than with revolution.

the activities of the Russian Social Democratic Party were themselves a reaction to the Russian autocratic state.

Lenin's hope was for a breakdown of the existing regime in Russia. His kind of revolution could only thrive on political chaos. In 1905 the conditions were not right. Despite the mutiny on the *Potemkin*, the sailors' mutiny at Kronstadt and revolt among the troops returning from the Russo–Japanese War on the Trans-Siberian Railway, these incidents had only a short-term effect and the autocracy, though shaken, remained stable, saved by Sergei Witte's October Manifesto, the creation of the Duma, and the overall loyalty of the majority of troops.

Of all the socialist groups, it was the Social Revolutionaries, with their support from the peasantry, who achieved the most. To some extent, this supports the view that the 1905 Revolution can be interpreted more as a series of rural revolts than an urban revolution, especially since the timing of urban unrest failed to coincide with rural unrest. Nevertheless, 1905 was the first time that unrest came from so many different quarters. The role of the SRs as successors to the populists (*Narodniki)* was confirmed by their agrarian support which they were able to maintain until 1917. Unlike the Bolsheviks, the SRs were more interested in rural issues than the development of a vast industrial proletariat in Russia. Their future ideas focused upon Russia as a huge peasant commune, composed of a mass of peasant village communities.

At the beginning of the 20th century, Russia was essentially an agricultural country. According to the 1897 census, out of a total population of 110 million nearly 97 million were peasants.

Despite their emancipation in 1861, and the abolition of redemption payments in 1905, the majority of peasants remained tied to their communes and could not travel freely. They had a separate legal status from the rest of the population, with their own law courts. Although output per capita, and even productivity per capita, had been growing from the last two decades of the 19th century until 1914, rural under-employment remained the main problem. Agriculture was becoming increasingly regionally differentiated, with further impoverishment at the centre, in contrast with expansion and improvement on the periphery. Ploughing was still largely carried out by a wooden cultivator rather than by an iron or steel plough share. One-third of all peasant holdings was without a horse and another third had only one horse. Harvesting was by hand, using sickles and scythes, rather than mechanical harvesters. There was a shortage of fertilizers and peasants often employed the strip system of cultivation, which seemed to be more rational economically under the prevailing conditions. Inefficient methods and subsistence farming were potentially dangerous, and could result in hardship and famine—such as those of 1891, 1897 and 1903 when many were on the verge of starvation. In turn, these conditions could lead to further agrarian unrest and the raiding of towns and country estates.

Leon Trotsky in prison after the 1905 Revolution had been suppressed. The St. Petersburg Soviet, of which Trotsky had been a member, had been the focus of much attention until its members were arrested en masse in December.

By 1914 agriculture was still the weakest part of the Russian economy, though there were signs of progress following reforms by ministers Witte and Stolypin, and a development in the use of agronomical methods. Some improvement came after 1906 with the introduction of Stolypin's reforms which aimed to create a large stratum of independent peasant farms, separated from the village communities, as a guarantee of political and social support for tsarism. This ended the earlier policy of supporting the peasant communes (*obschina* or *mir*) which the government had employed since 1861, as the peasant revolution in 1905–06 had been led by peasant communes who had undermined central authority. Yet by 1914 the majority of peasants still lived in peasant communes.

The ethnic make-up of the Empire provided another source of potential discontent due to the harassment of minorities, aggravated by the policy of Russification promoted in the late 19th century. Russia was divided into 96 provinces, each with a governor general who was the civil and military head and responsible to the Ministry of the Interior—and ultimately to the tsar. Most of the provinces were non-Russian, although all the governor generals were Russian. The Orthodox faith served as an additional instrument of Russification, having an especial impact upon the Catholics in Poland and Ukraine, and against Protestantism in the Baltic regions. Baltic Lutherans were forcibly converted to Orthodoxy on the pain of imprisonment. In Georgia the language was suppressed and enforced Russification was introduced. There was nationalist resistance against this, such as the *Messame Dassy*, one of whose members was the fifteen-year-old seminarist, Joseph Vissarionovich Djugashvili (Stalin). Meanwhile, in Central Asia there was less Russification. Among the Tatars about 10,000 were converted to Orthodoxy, but in most areas Islam remained strong.

Russification meant that Russian became the language of social advancement and that national languages were not recognised. Thus minority peoples could not speak their mother tongue in the schools or the administrations. Russia was not unique in this policy; parallels can be found in Britain and France at the time, with the treatment of Welsh and Breton respectively. Similarly, Bismarck's *Kulturkampf* had been imposed upon Germany's Polish possessions. But it was the intensity of the campaign in Russia which is of importance, especially when linked to the deep anti-semitism throughout the Empire, as witnessed by the occasional outbreak of *pogroms*, especially those of 1903, supported by Interior Minister Plehve.

Despite these repressive measures, in the last decade of peace fundamental economic and social reforms were gradually being introduced into Russia. Social and political evolution was taking place, even if all four Dumas were hamstrung by the Fundamental Laws whereby the Duma could not pass laws autonomously and could be prorogued at the tsar's wish. The *zemstvos* carried out an extension of the health services in the provinces. In 1908 compulsory and universal education within 10 years had become a declared aim. By 1914 the government had established 50,000 additional primary schools. Without the explosive impact of international affairs on Russia, the Romanov dynasty may well have survived, although probably in a more constitutional form.

When Archduke Franz Ferdinand was assassinated in Sarajevo by Bosnian Serb members of the *Crna Ruka,* the Austro-Hungarian government provoked a dispute with Serbia. Pan-Slavist sentiment led the Russian government to support their Serbian co-religionists. Austria-Hungary, encouraged by Germany, attacked Serbia. Russian mobilization against Austria began and Germany began to mobilize against Russia. The Triple Entente, although not an alliance, nevertheless led France and Britain to support Russia against the Central Powers; by August the Great War engulfed Europe.

Initially, the war was popular in Russia, as it was throughout the rest of Europe. Men rallied to the flag and grievances were forgotten. In August 1914 Russian troops advanced into East Prussia, but defeats at the hands of the Germans at Tannenburg and on the Masurian Lakes proved serious setbacks to the Russian cause. The heavy losses in men and artillery, which the

Russians could ill afford, allowed the Germans to push into Russian Poland, taking Warsaw in October 1914. It has been estimated that by the end of the year the Russian army had suffered 4 million casualties. However, on the positive side, the defeat at Tannenburg had saved Paris, since it relieved the pressure on the French army which had been beaten back to the Marne by the invading Germans. Herein lay one of the strategic reasons for Russia's failure; the Russian *stavka* (high command) could never act independently since Russian campaigns were always coordinated with those of their French and British allies. Thus, as the campaign of 1914 had saved Paris and the Marne, so Brusilov's offensive in 1916 would take German pressure off the western allies at Verdun and on the Somme.

When considering Russia's wartime record, the standard argument maintains that the Russian army was poorly-equipped and badly led, but there is a tendency to exaggerate the situation. It is true that in 1914 Russia was short of artillery pieces, rifles, machine guns and barbed wire, and that the hospital system and medical services were woefully inadequate, but withtime these problems were resolved satisfactorily. Admittedly, the Russians consistently suffered at the hands of the Germans, yet it must be remembered that Germany was one of the most highly industrialized states in Europe at the time. Other national armies suffered severely when confronting the German military machine, or when German troops were sent to stiffen the

General strike of weavers in Ivanovo-Voznesensk in 1905, painted by G.A. Kuyazhevsky, F.M. Kulagin and V.N. Govorov in 1955. The general breakdown of authority in 1905 resulted in widespread strikes and peasant riots across the empire.

On 9 January 1905 the Tsar's troops opened fire on a peaceful demonstration taking place near the St. Petersburg Winter Palace. Hundreds of demonstrators died and the incident introduced the term 'Bloody Sunday' to the Russian vocabulary. The leader of the procession, Father George Gapon, was lynched a year later when it was discovered that he had once been a police spy.

fighting spirit of their allies—as on the Isonzo front in 1917. However, against Austria-Hungary the situation was rather different and the Russian army tended to conduct itself well. Eyewitness accounts tell of the *élan* of Russian regiments as they occupied Ukrainian villages at the time of Brusilov's offensive. Their horses were fresh, the leatherwork highly polished and the troops looked fit and well-turned out by comparison with the slovenly appearance of the retreating Austro-Hungarian troops and the accusations of pillage and arson levelled against them.

Industrialization by 1900

An increasing awareness of the influence of economic efficiency on political status led to a reshaping of the face of Russia.

"The owners of several factories, taking advantage of their superior position, do not hesitate to violate the conditions agreed upon with their employees."
State Council, 1886

For a decade and a half from the mid-1880s, Russia witnessed an upsurge of largely state-initiated industrialization. This programme was guided by the Minister of Finance, Sergei Witte, and was driven primarily by the realization that power and prestige in Europe rested on military strength which, in turn, depended increasingly on economic modernization. This accounts for an industrialization strategy slanted toward heavy industry, linking defence industries with railway expansion.

The demands of rapid industrial development meant a concomitant expansion of the urban work-force; numerous towns and cities sprang up or expanded quickly to accommodate them. The urban population increased from around 10 per cent of the empire's total population (about 9 million) in 1867 to 21 per cent (25.84 million) in 1917. Of the towns, 36 possessed some 100,000 inhabitants each; by 1910, St. Petersburg and Moscow ranked among the 10 largest in Europe. Yet severe over-crowding and insanitary living conditions resulted in frequent outbreaks of diseases such as cholera and typhus. High rates of urban mortality led to most of this population increase stemming from peasants moving into the towns.

The expansion of the railways in the 1870s led to the creation of a national market, providing a link between the highly-populated central regions and southern Russia and, in particular, all year round access to the warm-water port of Odessa. The construction of the Trans-Siberian Railway in the decade after 1891 provided a military link to Asiatic Russia, but also facilitated economic exploitation of the region.

In European Russia, between 1906 and 1911, an attempt was made to modernize peasant agriculture by introducing a land reform which would replace the village commune with individual family farmsteads. Although much was claimed for the success of the reforms of Pyotr Stolypin, by 1914 less than 25 per cent of all peasant households had established such independent private farms. While the state continued to play an important role in industrialization after 1907, this latter phase of economic expansion was characterized by more private entrepreneurship and capital (both native and foreign). Russia also enjoyed a run of good harvests, and relatively high international wheat prices helped to produce a favourable balance of trade. Yet, on the eve of war, the benefits of such growth had not been felt by many. A low per capita income was joined to low average rates of literacy, and high rates of mortality (especially infant). Overall, much social discontent stemmed from the wide differences in wealth and life expectancy between and within social groups.

Oil derricks in Baku, Azerbaijan in the 1890s. During the period between 1898-1901, Russia was producing more oil than the rest of the world put together. Caucasian oil production, in particular, soared above all previous records.

Industry and agriculture

- — Russian border 1900
- ● major manufacturing centre
- ⚒ heavy industry
- ◎ textiles
- ▢ food processing
- ⛏ oil
- ⊨ railway
- ▨ area with greatest influx of workers
- ⚓ major trading port

NORWAY

SWEDEN

Tornio

Uleaborg

Umea

Vasa

Finland

Abo

Helsinki

Stockholm

Vyborg

Reval

Estland

Baltic Sea

Pernov

Courland

Libau

Riga

Kovno

Kovno

GERMANY

Vilna

Vilna

Grodno

Polish Provinces

Byalistok

Warsaw

Lvov

AUSTRO-HUNGARIAN EMPIRE

Carpathian Mts

Zhitomir

Volhynia

Jassi

Kishinev

Bessarabia

Braila

Izmail

Bucharest

Pleven

Ruse

Constanta

BULGARIA

Danube

Constantinople

Black Sea

Barents Sea

White Sea

Archangel

Archangel

N. Dvina

Vologda

60°

Lake Onega

Olonets

Kotlas

Vologda

Vologda

R U S S I A

Urals

Perm

Perm

Ekaterinburg

Viatka

Viatka

Viatka

St Petersburg

St Petersburg

Narva

Novgorod

Novgorod

Pskov

Pskov

F L A X

Tver

Yaroslavl

Yaroslavl

Kostroma

Kostroma

Nizhniy-Novgorod

Nizhniy-Novgorod

Kazan

Kazan

Ufa

Ufa

Volga

Vitebsk

Vitebsk

Smolensk

Smolensk

Moscow

Moscow

Vladimir

Simbirsk

Simbirsk

Mogilev

Mogilev

Kaluga

Kaluga

Ryazan

Penza

Penza

Samara

Orenburg

Orenburg

Minsk

Minsk

Bobruisk

Tula

Tula

Orel

Tambov

Tambov

Samara

Samara

R Y E

Orel

Chernigov

Chernigov

Kursk

Kursk

Voronezh

Voronezh

Saratov

Saratov

Uralsk

Uralsk

Kiev

Kiev

Poltava

Poltava

Dnieper

Kharkov

Kharkov

Pavlosk

Don

Podolia

Dniester

Ekaterinoslav

W H E A T

Donets

Don

Tsaritsyn

Kherson

Nikolaev

Taganrog

Novocherkassk

Don

Astrakhan

Astrakhan

Gurev

Ural

Emba

Odessa

Kherson

Kherson

Azov

Taurida

Kerch

Ekaterinodar

Stavropol

Stavropol

Caspian Sea

Trans-Caspia

Sevastopol

Novovossiisk

Kuban

Terek

Grozny

Poti

Batum

by 1878 to Russia

Trans-Caucasian Provinces

Baku

OTTOMAN EMPIRE

PERSIA

N

50°

40°

20°E 70°II

30°

40°

0 — 400 km

0 — 250 miles

The Russo–Japanese War, 1904–1905

Contempt for its Asiatic rivals led Russia to defeat and a loss of prestige in the world arena.

In the closing decades of the 19th century it became clear that any extension of Russian influence could not be sought in the European sphere. Hence the turn to the Far East, seemingly attractive because potential rivals were (foolishly) underrated; contempt was only shown towards the Japanese while it was believed a "peaceful penetration" of China could enable Russia to become the dominant naval power in the Pacific.

After China's defeat in its war with Japan in 1894–95, Russia's influence increased with the concession to build the Chinese Eastern Railway across Manchuria, followed in 1898 by another from Harbin to Port Arthur. Russia then turned its attention toward Korea, an area of great concern to the Japanese since it afforded the most direct access to the Chinese mainland. Although the Japanese were willing to negotiate a division of spheres of influence, the Russian government remained contemptuous.

Japan responded in February 1904 by launching a surprise torpedo-boat attack on the Russian Far Eastern fleet in Port Arthur, causing immense damage. The Russian government was taken by surprise and was left militarily unprepared. The Trans-Siberian railway was still incomplete and supplies to the Far East had thus to be transported via Lake Baikal. In the Far East Japanese troops outnumbered Russian forces and benefited from easy reinforcement by sea. Russia did not possess naval superiority and had only two widely distant naval bases—Port Arthur and Vladivostok—in comparison with Japan's numerous home ports.

A Japanese print reveals the effects of Japan's naval superiority during the Russo–Japanese War. The Russian ship pictured is the flagship Petro Pavlovsk *sunk on 13 April 1904 with the loss of over 800 seamen.*

A succession of Japanese victories, including the seizure of the port of Dairen, enabled them to besiege Port Arthur, which finally surrendered in December 1904 after a siege of 148 days. Russia was left with only one usable naval base in the Far East—Vladivostok. The land war culminated in the battle of Mukden, the largest land battle to date, in which the Tsar's forces were resoundingly, though not conclusively, beaten.

Meanwhile, the sea war continued to favour Japan. Having made an epic global voyage from Reval to the South China Sea, the Russian Baltic Fleet ran directly into the powerful fleet of Admiral Togo in the Straits of Tsushima on 27 May 1905. Within hours the entire Russian squadron had been destroyed.

In August 1905, the Treaty of Portsmouth was ratified, with an humiliated Russia recognizing Japan's interest in Korea and ceding to Japan the southern part of Sakhalin Island, control of the Liaotung Peninsula, together with Port Arthur and Dairen.

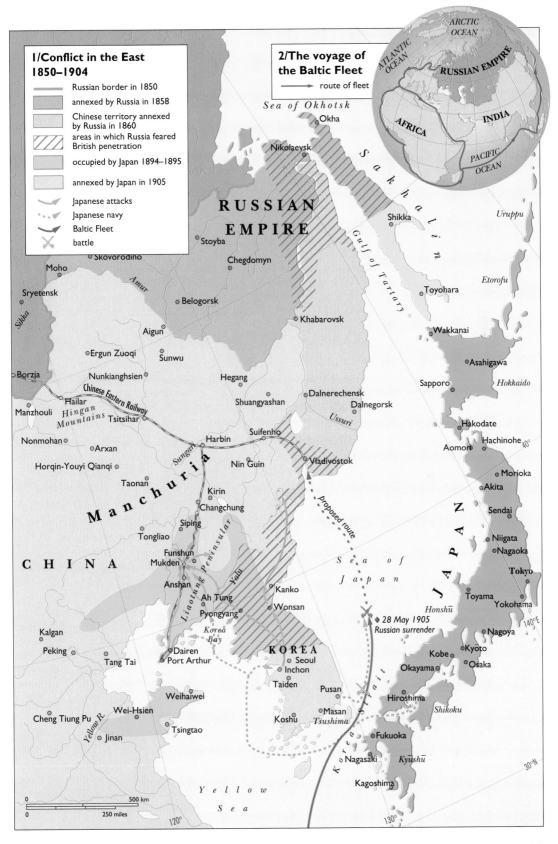

1/Conflict in the East 1850–1904

- Russian border in 1850
- annexed by Russia in 1858
- Chinese territory annexed by Russia in 1860
- areas in which Russia feared British penetration
- occupied by Japan 1894–1895
- annexed by Japan in 1905
- Japanese attacks
- Japanese navy
- Baltic Fleet
- battle

2/The voyage of the Baltic Fleet

→ route of fleet

ARCTIC OCEAN
ATLANTIC OCEAN
RUSSIAN EMPIRE
AFRICA
INDIA
PACIFIC OCEAN

Sea of Okhotsk
Okha
Nikolaevsk
Sakhalin
Gulf of Tartary
Shikka
Uruppu

RUSSIAN EMPIRE

Stoyba
Chegdomyn
Etorofu

Moho
Skovorodino
Amur
Toyohara

Sryetensk
Belogorsk
Wakkanai

Silka
Khabarovsk

Aigun
Asahigawa

Ergun Zuoqi
Sunwu
Sapporo
Hokkaido

Borzja
Nunkianghsien
Hegang
Dalnerechensk

Manzhouli
Hailar
Chinese Eastern Railway
Shuangyashan
Dalnegorsk
Hakodate

Hingan Mountains
Tsitsihar
Ussuri
Hachinohe
40°

Nonmohan
Harbin
Suifenho
Aomori

Arxan
Sungari
Nin Guin
Vladivostok
Morioka

Horqin-Youyi Qianqi
Akita

Taonan
Kirin
Sendai

Manchuria
Changchung
Niigata

Tongliao
Siping
Nagaoka

Funshun
Mukden
proposed route
Tokyo

CHINA
Anshan
Toyama

Ah Tung
Yalu
Kanko
Honshū
Yokohama

Kalgan
Pyongyang
Wonsan
Sea of Japan
Nagoya
140°E

Peking
Korea Bay
28 May 1905 Russian surrender
Kobe

Tang Tai
Dairen
Port Arthur
KOREA
Seoul
Inchon
Kyoto
Osaka

Taiden
Pusan
Okayama

Weihaiwei
Hiroshima
Shikoku

Cheng Tiung Pu
Wei-Hsien
Koshu
Masan
Tsushima

Yellow R.
Jinan
Tsingtao
Fukuoka

Nagasaki
Kyūshū
30°N

Kagoshima

Korea Strait

Yellow Sea

0   500 km
0   250 miles

120°   130°

The 1905 Revolution

Born of widespread discontent and encouraged by political agitators, the 1905 Revolution pushed Nicholas II toward constitutional monarchy.

"Unrest and disturbances in the capitals and in many parts of Our empire fill Our hearts with great and heavy grief. The welfare of the Russian sovereign is inseparable from the welfare of the people, and the people's sorrow is His sorrow."

Nicholas II, 17 October 1905

The year 1905 witnessed Russia's first real, though ill-fated, revolution. Essentially a spontaneous and unconcerted expression of discontent, the Revolution resulted from a number of factors.

The humiliating defeat suffered in the war with Japan undoubtedly provided the short term fuse, but the Revolution was the fruit born of a much wider malaise. While inadequacies of the emancipation settlement had produced discontent among the peasantry, this was exacerbated by the "agrarian crisis" at the end of the 19th century when an already land-hungry peasantry was partly depressed by the burdens of industrialization. The peasants continued to demand land—principally that belonging to gentry landlords—which they believed to be rightfully theirs. An agrarian revolution of enormous intensity broke out around the agricultural centre in 1905–06. Although characterized by local particularism, a national Peasants' Union was also created—partially as a result of the efforts of the Socialist Revolutionary Party.

Dissatisfaction was also rife in the urban centres, where living and working conditions were appalling and wages both low and irregular. Such conditions were ideal for exploitation by revolutionaries and their propaganda, but although these had the desired effect, the urban and rural revolutions failed to coincide—a factor that proved crucial to the survival of tsarism.

Another vital factor in the Tsar's survival was the fact that, despite some small-scale mutinies, including the seizure of the battleship *Potemkin*, the armed forces remained loyal. The revolution was, in fact, largely sparked by the actions of the army when they fired upon a peaceful demonstration led by the priest, Father George Gapon, on 9 January 1905.

The most notable action of the revolutionary groups during 1905 was the attempted seizure of power in St. Petersburg during December. The Petrograd Soviet, led by Trotsky and providing the model for that of 1917, collapsed when martial law was declared and all meetings forbidden. Large areas of Moscow were also seized but by failing to seize Nikolaevsk Station, the revolutionaries left themselves open to attack by government troops and the uprising was suppressed.

Though Nicholas II was once more firmly in control by 1906, the 1905 Revolution did achieve at least one significant result: the establishment of the State Duma, a creation which many saw as the first step down the path toward democratic government.

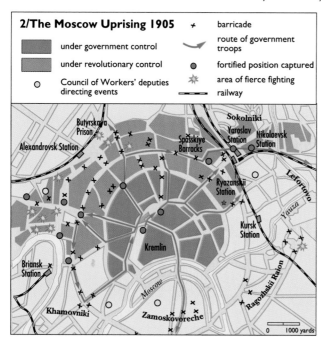

2/The Moscow Uprising 1905

- under government control
- under revolutionary control
- ○ Council of Workers' deputies directing events
- ✷ barricade
- route of government troops
- ● fortified position captured
- ✻ area of fierce fighting
- railway

Butyrskaya Prison
Alexandrovsk Station
Sokolniki
Spasskiye Barracks
Yaroslav Station
Nikolaevsk Station
Ryazanskii Station
Lefortovo
Yauza
Kursk Station
Kremlin
Briansk Station
Ragozhskii Raion
Moscow
Khamovniki
Zamoskvoreche

0   1000 yards

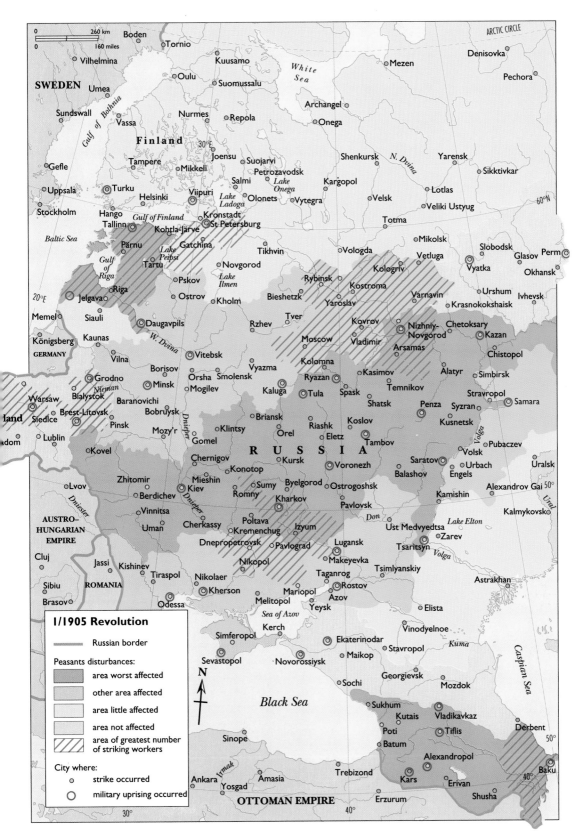

0   260 km
0   160 miles

ARCTIC CIRCLE

SWEDEN

Boden
Tornio
Kuusamo
Vilhelmina
Mezen
Denisovka
Pechora
Oulu
Umea
Suomussalu
White Sea
Sundswall
Gulf of Bothnia
Nurmes
Repola
Archangel
Yarensk
Vassa
Onega
Sikktivkar
Gefle
Finland
30°E
Joensu
Shenkursk
N. Dvina
Tampere
Suojarvi
Petrozavodsk
Lotlas
Mikkeli
Salmi
Lake Onega
Kargopol
Veliki Ustyug
60°N
Uppsala
Turku
Helsinki
Viipuri
Olonets
Velsk
Okhansk
Stockholm
Hango
Lake Ladoga
Vytegra
Totma
Gulf of Finland
Kronstadt
Mikolsk
Slobodsk
Tallinn
Kohtla-Järve
St Petersburg
Vologda
Vetluga
Glasov
Perm
Baltic Sea
Pärnu
Gatchina
Tikhvin
Kologriv
Vyatka
Gulf of Riga
Lake Peipsi
Novgorod
Varnavin
Urshum
Ivhevsk
20°E
Tartu
Lake Ilmen
Rybinsk
Kostroma
Krasnokokshaisk
Riga
Pskov
Bieshetzk
Yaroslav
Jelgava
Ostrov
Kholm
Tver
Kovrov
Chetoksary
Memel
Siauli
Daugavpils
Rzhev
Nizhniy-Novgorod
Kazan
Königsberg
Kaunas
W. Dvina
Moscow
Vladimir
Arsamas
Chistopol
GERMANY
Vilna
Vitebsk
Vyazma
Kasimov
Alatyr
Simbirsk
Borisov
Orsha Smolensk
Kolomna
Temnikov
Stravropol
Grodno
Nieman
Minsk
Mogilev
Kaluga
Ryazan
Tula
Shatsk
Penza
Syzran
Samara
Warsaw
Bialystok
Baranovichi
Spask
Kusnetsk
land
Siedlce
Brest-Litovsk
Bobruysk
Briansk
Koslov
Tambov
Pubaczev
dom
Lublin
Pinsk
Mozy'r
Gomel
Klintsy
Orel
Eletz
Volsk
Uralsk
Kovel
Chernigov
Kursk
Riashk
Saratov
Urbach
Alexandrov Gai
50°
Lvov
Zhitomir
Konotop
Voronezh
Balashov
Engels
Ural
AUSTRO-HUNGARIAN EMPIRE
Berdichev
Mieshin
Kiev
Sumy
Byelgorod
Ostrogoshsk
Kamishin
Kalmykovsk
Vinnitsa
Romny
Kharkov
Pavlovsk
Cluj
Uman
Cherkassy
Poltava
Izyum
Don
Ust Medvyedtsa
Lake Elton
Jassi
Kishinev
Kremenchug
Pavlograd
Lugansk
Tsaritsyn
Zarev
Volga
ROMANIA
Tiraspol
Nikolaer
Nikopol
Makeyevka
Tsimlyanskiy
Astrakhan
Sibiu
Brasov
Kherson
Melitopol
Mariopol
Taganrog
Rostov
Azov
Elista
Odessa
Yeysk
Sea of Azov
Kerch
Vinodyelnoe
Simferopol
Ekaterinodar
Stavropol
Kuma
Caspian Sea
Sevastopol
Novorossiysk
Maikop
Georgievsk
Mozdok
Sochi
Black Sea
Sukhum
Kutais
Vladikavkaz
Derbent
Poti
Tiflis
50°
Sinope
Batum
Alexandropol
Baku
40°
Ankara
Yosgad
Amasia
Trebizond
Kars
Erivan
Shusha
OTTOMAN EMPIRE
Erzurum
30°
40°

I/1905 Revolution

——— Russian border

Peasants disturbances:

area worst affected

other area affected

area little affected

area not affected

area of greatest number of striking workers

City where:

o   strike occurred

O   military uprising occurred

N

World War I

The war in Russia was popular at first. Men rallied to the flag and for a while social and political grievances were allayed.

"Thing are getting worse; the men are splendid, there are plenty of guns and ammunition, but there is a lack of grey matter in the generals' heads … We are ready to die for Russia, but not for the whim of a general."
Lt. Rodzyanko, son of the President of the Duma, 1916

As the once powerful Ottoman Empire slowly declined, the European powers struggled to gain control of the Balkan peninsula with its access to the Near East. Turkish atrocities against Bulgarians in the 1870s provided Russia with a pretext for declaring war on Turkey. Following Russia's victory, the Treaty of San Stefano required Turkey to agree to the creation of the independent state of Bulgaria. Britain, Germany and Austria-Hungary, however, over-ruled this plan at the Congress of Berlin; a much smaller Bulgaria was agreed, with adjacent territory being given to Romania, Serbia, Montenegro and Greece.

By 1914 Serbia had become Russia's main Balkan ally. When, as a result of the assassination of Archduke Franz Ferdinand by a Serbian nationalist, Austria-Hungary threatened Serbian independence, Russia had no option but to go to her ally's aid. Germany chose this as a pretext for declaring war on Russia, then France. Great Britain was drawn into the conflict and Russia's commitment was sealed by its treaty obligations with its allies.

In August 1914 Russian troops advanced into East Prussia, only to be defeated with heavy losses at Tannenburg and on the Masurian Lakes. The German counter-attack penetrated into Russian Poland, taking Warsaw in October 1914. Throughout 1915, Russia gradually relinquished more territory along an 800-mile front.

However, the Russian defeat at Tannenburg had saved Paris by relieving pressure on the French army which had been beaten back to the River Marne. Herein lay one of the reasons for subsequent Russian set-backs: its campaigns had always to be coordinated with the needs of its allies. Thus, in the summer of 1916, Brusilov's offensive took German pressure off the Allies at Verdun, the Somme and the Italian battles on the Isonzo front. But between the autumn of 1916 and the events of February/March 1917, the Russian army was forced into an exhausting defence, compounded by heavy losses, desertions and declining morale due to growing concern about the deteriorating situation at home.

Russian infantry parade in full kit during the summer of 1914.

Russia and World War I

boundaries in 1914

Entente Powers and their allies

Central Powers and their allies

neutral

major Russian offensive

Front lines:
Dec 1914
Nov 1915
Dec 1916
Dec 1917

major battle

area of mutiny of Russian army July 1917

ARCTIC CIRCLE

0   500 km
0   300 miles

N

70°

60°

10°

50°

40°N

20°E

30°

40°

N O R W A Y

S W E D E N

Oslo

Stockholm

Gulf of Bothnia

Finland

White Sea

Murmansk

Lake Onega

Lake Ladoga

Helsinki

Gulf of Finland

Tallinn

Estonia

Lake Peipus

Livonia

Gulf of Riga

Baltic Sea

Courland

Riga

Königsberg

Elbing

East Prussia

Tannenberg

GERMANY

Lithuania

Gumbinnen

Vilna

Vilkovisky

Masurian Lakes

Niemen

W. Dvina

Petrograd
St. Petersburg

Vologda

Viatka

Yaroslavl

Tver

Nizhniy-Novgorod

Moscow

Smolensk

Ryazan

Minsk

Mogilev

Belorussia

Tula

Penza

Lodz

Warsaw

Brest-Litovsk

Poland

Pinsk

Gomel

Orel

Vistula

Oder

Krasnik

Komarov

Pripet

Marshes

Dnieper

R U S S I A N   E M P I R E

Volga

Lemberg

Dniester

Kiev

Voronezh

AUSTRO-HUNGARIAN EMPIRE

U k r a i n e

Kharkov

Don

Tsaritsyn

Volga

Ural

Ekaterinoslav

Danube

Bessarabia

Moldavia

Nikolayev

Rostov

Don

Astrakhan

ROMANIA

Ismail

Odessa

Crimea

Sea of Azov

Kerch

Novovosissk

Caspian Sea

SERBIA

MONTE-NEGRO

Bucharest

Danube

Sevastopol

Black Sea

50°

BULGARIA

ALBANIA

GREECE

Constantinople
Istanbul

Sinope

Batumi

Tiflis

Gallipoli

Trebizond

Georgia

Kars

Azerbaijan

Aegean Sea

O T T O M A N   E M P I R E

Tabriz

to Italy

PERSIA

91

The Tsars

With Ivan the Terrible began a system of autocracy in Russia which would last until 1918, when its last representative was shot and bayonetted in a cellar in Ekaterinburg.

"The intention and the end of monarchy is the glory of the citizens, of the state, and of the sovereign."
Catherine II, in her *Instructions*, 1767

Ivan the "Terrible" (1533–84)—or more accurately "the Dread" (*grozny*)—called himself not just Tsar, thus giving greater prestige to the Muscovite ruler than the earlier title of Grand Prince, but Tsar "of all Russia" (indicating Moscow's authority over all Russia). An awesome and tyrannical figure who dominated over half the 16th century, his paranoia and excessively cruel and sadistic behaviour have been explained by events in his childhood, and reinforced by the belief that his wife Anastasia had been poisoned. Some twenty years later he was to strike dead his son (and heir), causing his second wife to miscarry, further aggravating his mood swings and unpredictable rages.

Tsardom moved into a new phase in 1613 when Mikhail Romanov brought an end to the Time of Troubles and established the dynasty that was to last some 300 years. Within a century of Ivan there were two "Greats"—Peter I and Catherine II, both in their own ways deemed modernizers and westernizers by historians. Peter created the new capital of St. Petersburg—giving him and Russia a "window to the west"—and instituted industrial development, though largely to suit his military purposes. While Peter was a reformer, his reforms were geared to suit the political, administrative and fiscal needs of the state, and his economic measures ironically acted to strengthen serfdom. These seeming paradoxes were matched by his personal life. He was "a giant of a man" (with a preference for low ceilings and the company of dwarfs and giants), and a ruler's reputation to match (sacrificing thousands, for instance, to the construction of St. Petersburg). Following his motto, "I am a student and I seek teachers", he disguised himself as a labourer to work in the shipyards of Amsterdam and Greenwich; indulged in dentistry; carried out exceedingly dangerous experiments with fireworks; relished watching operations; performed autopsies; made his courtiers wear western clothes and shave off their beards; forced guests at gunpoint to drink copious quantities of vodka and tore off the nose of an Egyptian mummy in a fit of pique.

The reign of Anne (1730–40) was more subdued, renowned more for its lavish court and the ascendancy of the "German party" which was deposed when Peter's daughter Elizabeth took the throne. The latter, characterized as "good natured, indolent and extravagant", allegedly had 15,000 dresses and left the Winter Palace uncompleted on her death. She believed reading was an unhealthy occupation, and that Britain was joined to the Continent. Westernization continued in the arts (which she patronized) though German influence gradually gave way to French. While Catherine's reign (1762–96) began brutally—she deposed her husband by armed insurrection—it contained both colour and wisdom: she was considered exceedingly beautiful and had a long succession of lovers. She was also intelligent. As an enlightened autocrat, influenced by the political literature of the time, she believed that the purpose of government was to mobilize resources in the interests of the welfare of the people and the power of the state, though society would remain divided into social "estates" (nobles, peasants, urban dwellers). Her reign saw financial and administrative reform, educational measures and religious tolerance.

After the reigns of two reactionary tsars, Alexander I (1801–25), who feared revolution spreading from Europe, and Nicholas I (1825–55), who installed secret police to protect autocracy, another reformer, Alexander II, was to abolish serfdom. This "liberator" tsar introduced a number of reforms—in education, local government, the legal system and the military—which set the path for change. But these reforms highlighted the dilemma for tsarism in the modernizing world: to allow a modicum of change to appease the people but not so much that it undermined tsarism. Reform, paradoxically, was intended to preserve the status quo and consolidate autocracy. The final irony was that Alexander was assassinated by the very forces that had been set in motion by his reforms.

The reign of his son, Alexander III, horrified by the carnage of his father's death, saw the pendulum swing back towards repression. Nicholas II, who brought tsardom into the 20th century, was very much a man out of time, ill-suited to his role as tsar and with no desire to assume the mantle. A product of a reactionary upbringing, he became tsar in the midst of industrialization—a process at odds with autocracy (implemented to maintain Russia's international political position and prestige)—and within years was countering the 1905 revolution. The concessions granted as a result and their half-hearted implementation once again revealed the tsarist dilemma over reform. Nicholas withdrew into his family, becoming convinced of the healing powers of the peasant holy-man, Rasputin, over his haemophiliac son Alexis. After Russia had become embroiled in World War I, Nicholas only aggravated matters by assuming command himself, thus identifying the war's disasters with the tsar personally. When he abdicated in 1917, faith in tsardom was waning. The death of the entire family at the hands of the Bolsheviks in 1918 was to remain a poignant revolutionary symbol for the 20th century. The Romanovs, born out of one time of troubles, were destroyed in the midst of another.

The coronation of Nicholas II, painted by Henri Gervex. Beginning amidst all the pomp and ceremony traditionally expected of a tsar's accession, Nicholas' reign was to end in the cellar of a merchant's house in Ekaterinburg. Unable to empathize with the plight of his people and unwilling to relinquish any vestige of his autocratic power, Nicholas, along with all his immediate family, was slaughtered on the orders of Russia's new dictator, Lenin.

VI: From Revolution to the USSR

Although initially able to support the war effort, the economy was at breaking point when civilian unrest blazed into revolution, providing ideal conditions for exploitation by the Bolsheviks.

> *"The soldiers, workers and peasants did not overthrow the government of the Tsar and Kerensky merely to become the cannon fodder of the Allied imperialists."*
> Leon Trotsky

The Russian economy could not sustain a long drawn out war; the longer the conflict continued, the more the defects within the Russian economy would be revealed and social tensions accentuated. Military setbacks on the Eastern Front did not in themselves usher in the February Revolution. The war was not, as yet, unpopular. Generally speaking, the Russian army acquitted itself reasonably well, at least in its operations against the Austro-Hungarian Army. Deteriorating economic conditions lay at the heart of Russia's difficulties. As industry was directed to the military needs of the war, output and trade were seriously disrupted. The inefficient use of railways for military purposes led to overloading and under-maintenance, which in turn affected the civilian economy since less agricultural produce was transported into the cities. Food retail decreased. Inflation grew and bread queues lengthened as workers, their families and garrison troops went hungry. The situation was exacerbated by profiteering, corruption and government instability. During two-and-a-half years of war, Russia had four prime ministers, three foreign ministers, three defence ministers and six Ministers of the Interior. Rumours of conspiracies and pro-German sympathies in court circles led to a growing distaste for the autocracy. In the meantime a wave of strikes broke out in the winter of 1916/17.

Although the events of February 1917 were initially brought about by industrial unrest, it was the soldiers' refusal to fire on the crowds which turned the demonstrations into a revolt. Ultimately, it was the aristocracy that

In the early hours of the morning of 17 July 1918, the tsar and tsarina with their five children were taken to the cellar of the Ekaterinburg house in which they had been imprisoned since May. There they were executed by members of the Cheka. The bodies were dismembered, burned, drenched with acid and thrown down a disused mineshaft.

A communist propaganda painting shows Lenin addressing a crowd of cheering workers. Any claim that Lenin's Bolshevik takeover was an expression of the people's will bore little relation to the truth. Many saw the events of October 1917 as a counter-revolution and their feelings were given violent expression in August 1918 when Fanya Kaplan, a Socialist Revolutionary, shot and seriously injured Lenin.

brought about the abdication of the tsar. The liberals of the Duma appointed a committee to take temporary control of the governing of the country. Established in the Tauride Palace, this Provisional Government promised to hold elections, based upon universal suffrage, for a Constituent Assembly.

Russia remained in the war in the hope of achieving an all-out victory and winning back the territory it had lost to the Central Powers. At the same time the Provisional Government, fearing widespread agrarian unrest, needed the support of both the Allies and its own generals. The decision to continue the war meant delays in the reform process which ultimately undermined the authority of the Provisional Government.

Led by the *Zemstvo* leader, Prince Lvov, the ten-man administration of the Provisional Government was composed of Milyukov' Consitutional Democrats (Kadets), Gushkov (Octobrist) and Kerensky (SR). From the beginning, it had to share power with another political force which arose from the ashes of its prohibition in 1905; made up of workers' and soldiers' representatives, Socialist Revolutionaries and Mensheviks, the Petrograd Soviet established itself in another wing of the Tauride Palace and acted as a counter-balance and counter-signatory to the wishes and decrees of the Provisional Government.

The Provisional Government immediately began to establish a number of democratic institutions and implemented a wide range of civil liberties and reforms, such as the abolition of the death penalty. In the meantime, on 1 March the Petrograd Soviet had issued Order No. 1 in an attempt to put pressure on the Provisional Government. This would be blamed later for playing a considerable part in the subsequent breakdown of discipline in the Russian armed forces.

Meanwhile, the war continued. The Germans, hoping that Lenin's return to Russia would lead to unrest and an end to Russia's war effort—and consequently to the release of German troops from the Eastern Front—allowed Lenin and his followers to return to Russia on a special sealed train which travelled through Germany, across the Baltic, through Sweden, into Finland and from there into Russia.

In May, Alexander Kerensky, a leading member of the Petrograd Soviet, was made Minister of War. Kerensky's aim, like that of many others in the Provisional Government, was to delay major reforms until victory over Germany had been achieved. In July, he mobilized the troops in a last great offensive against the Germans. But he had miscalculated the wishes of the Russian people—who had grown tired of war. The failure of the Kerensky offensive discredited the government and gave the Bolsheviks the chance to increase their support, but an abortive attempt by the Bolsheviks to seize power in July weakened their influence. Lenin was forced to seek refuge in Finland for three months, where he prepared his *State and Revolution.*

In the meantime, the Germans entered Riga. As Russian troops continued to desert, and more troublesome units were disbanded, German forces threatened Petrograd. Conservative elements began to lose faith in Kerensky and the Provisional Government. They found a new champion in General Kornilov, who in August marched on Petrograd to restore order. Kerensky, fearing Kornilov would take power, turned to the Bolsheviks who agreed to help. The capital was defended by a workers' militia, assisted by the sailors from the Kronstadt naval base and "Red Guards", who eventually captured Kornilov.

In the ensuing climate of anarchy, it became clear that the Provisional Government could neither continue the war nor finish it. With growing support, Lenin and the Bolsheviks overthrew the Provisional Government and in October 1917 seized power in Petrograd with relatively little loss of life.

Throughout 1917, the Bolsheviks had criticised the Provisional Government for failing to implement a Constituent Assembly. Yet the idea of establishing a democratically elected parliament ran counter to the Leninist belief, developed in 1902, that a single party should govern and administer alone. Nevertheless, if the Bolshevik government was to be fully legitimized, then some sort of compromise had to be made. Elections were held in November, but the results confirmed the Bolsheviks' worst fears: that their rivals, the SRs, were in the legitimate position for assuming power. With only 25 per cent of the vote, the Bolsheviks found themselves in the minority and had to share power with the Left SRs. The Constituent Assembly opened on 18 January; debating lasted for one day. When the delegates arrived the following day, Lenin had already posted his Latvian guards around the Tauride Palace, proroguing the first and only session of the Assembly. Russia had experienced just one day of "revolutionary democracy". This explains why Lenin's detractors referred to the October Revolution as the Counter-Revolution.

Lenin had clearly demonstrated that he was not prepared to reach a compromise with the other parties, which in his eyes could only lead to "... hesitations, impotence and chaos". In the continuing process of consolidating power, the Bolsheviks now turned their attentions to Russia's external threats. The war was still going on, and there was growing concern over the German armies that had penetrated deep into Russian territory. In March

1918, peace proposals were negotiated between Russia and Germany at Brest-Litovsk. As a result, Russia lost nearly a third of its agricultural land, half its industry, four-fifths of its mines and a third of its population. Brest-Litovsk was a "humiliating peace" (Trotsky), but it provided the Bolsheviks with a necessary breathing space to consolidate the revolution and to tackle the next problem: Russia's internal and external enemies in both the Civil War and Russo–Polish War. As Lenin put it: "We gained a little time and sacrificed a great deal of space for it." He still believed that the losses would soon be recuperated once workers throughout the rest of Europe had staged their own revolutions. As it was, by 1940 the Soviet Union would regain all territories lost at Brest-Litovsk.

The Civil War was also a war of foreign intervention. Before November 1918 the Allies were maintaining troops in Russia in an attempt to contain the Germans in the East. After the end of the World War I it became evident that their policy had been transmogrified into one of containing Bolshevism within Russia.

A Soviet propaganda poster of 1920 outlines the need for industrial expansion and proclaims the benefits of Lenin's New Economic Policy.

Hampered by quarrels, the counter-revolutionary White Armies of Kolchak, Yudenich, Wrangel and Denikin failed because they were separated by vast distances, precluding any concerted action whilst the Reds held the Russian heartland. Furthermore, the Whites consistently failed to win support in the areas they occupied. The peasants did not want a return to landlordism and feared retribution, as in 1905, neither were they prepared to tolerate requisitioning on the part of any "outside" forces. Ultimately, the utility of Western intervention was affected by war-weariness and poor morale. Some Allied soldiers even expressed more or less outspoken sympathy for the workers' government in Moscow.

The year 1921 was one of crisis, Soviet Russia, or the RSFSR, witnessing internal hostility to the regime, with sporadic peasant disturbances and riots in Petrograd and Kronstadt. Russia needed time to restore economic prosperity after the ravages of World War I and the Civil War. Peasant unrest was gathering momentum from the summer of 1920, and by the winter of 1920/21 was accompanied by widespread urban discontent. Factions in support of "workers" and against the centralisation and lack of Soviet democracy of the past few years had also emerged in the Bolshevik Party. In the spring of 1921 Kronstadt, the centre of Bolshevism in 1917, rebelled. The Bolsheviks, for their own survival, were forced to change course and adopted a series of measures that came to be known as the New Economic Policy. Government-regulated "War Communism", and the requisitioning of produce from the peasants, was replaced by a mixed economy. At the same time Lenin believed that there would have to be a tightening of political control and banned all factions in the Party. Party unity was to be the order of the day. At the Party Congress in March 1921, Lenin made it clear that the Party could no longer afford the luxury of factional debate, and the Central Committee could now expel anyone from the Party, including members of the Central Committee itself. The Cheka, or secret police, had already been formed early in 1918 under the leadership of Felix Dzerzhinsky. The Cheka arrested political opponents during the Civil War. Concentration camps appeared towards the end of 1918, and more than 12,000 people were executed between 1918 and 1921.

After Lenin's death on 21 January 1924, opposition to Trotsky led to a struggle for the overall control of the Party, with Stalin acting quietly behind the scenes, like a Machiavellian puppet master. To aid him in his work Stalin, as Party Secretary, maintained a filing system which contained details of all the foibles, mistakes and indiscretions of his fellow party members which he could use to his advantage in eliminating opposition. Stalin also drew upon his ability to form alliances with opposition groups who then helped him to destroy certain personalities. With the help of Kamenev and Zinoviev he had discredited Trotsky. He then turned on Kamenev and Zinoviev, and finally upon Bukharin. By the late 1930s, none of the original leaders of the Bolshevik Revolution had escaped execution.

Once in power, Stalin secured his position through ruthless policies of industrialization, collectivization and a system of terror and repression that grew with the spread of the Gulag system of "corrective labour"—linked with the Great Purges ,or *Yezhovshchina*, of 1936–41, during which it has been estimated that ten million were arrested and three million executed.

One particular group that suffered the effects of both the Gulag and collectivization was that of the alleged kulaks. The term "kulak" was a pejorative, catch-all phrase, with original connotations of usury. Lenin believed that Stolypin's agrarian reforms in 1906 had created wealthy peasants who acquired larger farms, yet they were not wealthy by western standards. Perceived by the Bolsheviks as a threat to the Communist state, and as petit-bourgeois, the Bolsheviks felt that they could not be relied upon to build

Lenin inspects troops of the universal military training courses on Red Square, 25 May 1919.

socialism. The NEP had generated capitalism and a market orientated private peasantry. Dubbed "New Exploitation of the Proletariat" by its detractors, NEP was resented by many Communists who saw kulaks and merchants getting richer whilst many workers remained poor. Recent data, from the opening up of Soviet archives, suggests that over a million peasant households (some five to six million peasants) were the victims of dekulakisation in 1930–33; another eight to nine million perished as a result of the 1932–33 famine and in the late 1930s possibly another one million were deported to less hospitable parts of the Soviet Union, where they suffered great hardship, many dying of destitution and brutality in the camps.

By the end of the 1930s, the Soviet Union had been through a tremendous upheaval. Between 17 and 18 million people had been lost in World War I and the Civil War, and a large number through famine and disease. A repressive totalitarian system had been put in place to maintain in power a party which had no legitimate democratic authority. Millions of citizens had been arrested and sent to special prison camps in the chaotic period of rapid industrialization and repressive collectivization. As international tensions mounted, an even more ferocious war was about to be unleashed on the Soviet people.

The New Moscow, *painted in 1937 by Yuri Pimenor. The vision of a thriving and happy society was carefully fostered by Stalin and his army of publicists. Foreign visitors were taken on strictly choreographed tours of the "booming" Soviet Union and they were encouraged to dispel rumours of oppression and hardship.*

1917 Revolutions

As the war continued, increasing economic pressure resulted in conditions ripe for a more radical solution to Russia's problems.

"Soldiers, Workers, Employees! The fate of the Revolution and democratic peace is in your hands."
Lenin (The Second All-Russian Congress of Soviets and Soldiers' Deputies)

A number of strikes broke out in Petrograd during January and February 1917. On 23 February, thousands of workers came out onto the streets, to be supported by a mutiny of the Petrograd garrison on 26 February. The uprising was spontaneous. There were no recognizable leaders, the revolt was not organized and there was no evidence of a conspiracy. It was the soldiers who turned riots into revolution when they refused to open fire on the rioters and instead distributed weapons to the students and workers.

On 27 February the Duma rejected the Tsar's decree to prorogue itself. They formed a provisional committee and hoped to persuade the Tsar to replace the autocracy with a constitutional monarchy. However the Tsar was abandoned by the politico-military élite, and abdicated at Pskov on 2 March. Banished to Tobolsk in Siberia and later to Ekaterinburg, the Tsar and his family were shot in July 1918.

Power was shared between Prince Lvov's Provisional Government and the Petrograd Soviet of Workers' and Peoples' Deputies. In the eight months that followed, the Provisional Government introduced a number of reforms, but it was hampered by its desire to continue the war with Germany, which grew more and more unpopular as the year progressed.

The Bolsheviks played no role in the February revolution. Lenin returned to Petrograd from Switzerland in a sealed train in April, but following an abortive attempt to seize power on 3 July, he was forced into exile again.

The Provisional Government was ultimately discredited by the failure of the Kerensky offensive in June and in early September by an unsuccessful attempt by the Commander-in-Chief, General Kornilov, to seize power. By the end of summer, popular support for the Bolsheviks was growing rapidly in the key urban centres as they gained the majority in the Petrograd and Moscow Soviets. Lenin returned to Russia on 10 October, as Kerensky, Lvov's successor faced growing opposition from the Petrograd Soviet.

Lenin felt that the time was ripe to seize power, although not all of the Bolsheviks, especially Kamenev and Zinoviev, were convinced by his arguments. Early in the morning of 24 October Red Guards began to take over the key points of Petrograd. On 25 October Red Guards attacked the Winter Palace, where the Provisional Government was in session, and arrested all the deputies, with the exception of Kerensky who fled the capital. In the meantime the All-Russian Congress of Soviets assembled at the Smolny Institute. The Bolsheviks' position was strengthened when Menshevik and Socialist Revolutionary demands for a left-wing coalition were rejected. On 26 October the Soviets entrusted authority to the Bolsheviks, who organised a Council of Peoples' Commissars as the executive body.

The Bolshevik Revolution witnessed little violence in Petrograd, and the next day most people went to work as usual without realising what had happened. Moscow fell into Bolshevik hands on 2 November, by which time most of the major industrial cities recognised Soviet power. The Bolsheviks arranged a cease-fire with Germany and turned their attentions toward consolidating power. Meanwhile anti-Bolshevik forces were preparing for battle.

Soldiers of the mutinous Petrograd garrison pose with their artillery pieces in front of the makeshift barricades. It was the role played by the troops which made a relatively unimportant civil disturbance into full-blooded revolution.

The Events of 1917

Russian territory under German military control in Feb 1917

Ukrainian People's Republic July 1917

Eastern Front 1914–1917

demoralized troops leave war zones, some living in "bands" behind the lines

★ principal strike

● Bolshevik controlled (strike)

route of troops sent against strikers

railways largely controlled by anti-Tsarist railway workers

route of Tsar's train

Lenin's route by rail

◇ 1916
strike in munitions factory suppressed by military force

◇ 23–26 Feb 1917
strikes
27 Feb 1917
troop mutinies
1917 October Revolution: Bolsheviks seize power

◇ Aug 1917
factory workers strike; demand end to war

ARCTIC CIRCLE

White Sea

Archangel

Pinega

N. Dvina

Lenin travels from Switzerland via Germany, Sweden and Finland ◇

SWEDEN

Hango

Lake Ladoga

Vyborg

Petrozavodsk

Lake Onega

RUSSIA

to Ekaterinburg

Baltic Sea

Reval

ESTONIA

Lake Peipus

Petrograd

Novgorod

Lake Ilmen

Malaya Vishera ◇ 1 March 1917
Tsar's train stopped

Vologda

Viatka

Viatka

LATVIA

Pskov

Dno

Staraya Russa

Bologoye

Yaroslavl

2 March 1917 ◇
Tsar abdicates
8 March 1917
Tsar arrested

Riga

Rzhev

Tver

Nizhniy-Novgorod

Kazan

Lithuania

Dvinsk

W. Dwina

Vitebsk

Moscow

Oka

GERMAN EMPIRE

Belorussia

Borisov

Orsha

Smolensk

Kaluga

1916–1917
Tsar's military HQ ◇

Bialystok

Nieman

Minsk

Mogilev

Tula

22 Feb 1917 ◇
Tsar's train leaves
from Mogilev

Penza

Warsaw

Brest-Litovsk

Pinsk

Pripet

Dnieper

Desna

Gomel

Orel

Tambov

Volga

Poland

Lublin

Saratov

July 1917 ◇
Mutinies. Hundreds of fleeing troops shot by government order

Dniester

Kiev

Voronezh

Slovakia

Tarnopol

16 July 1917 ◇
Ukrainian People's Republic proclaimed.

Poltava

Kharkov

Don

Lake Eltan

Tsaritsyn

AUSTRO-HUNGARIAN EMPIRE

Donets

Donbass

Volga

Prut

Dniester

Moldovia

Kishinev

Dnepropetrovsk

Novocherkask

Rostov

Astrakhan

Kherson

ROMANIA

Bucharest

Odessa

Sea of Azov

Crimea

Simferopol'

Kuma

Caspian Sea

Danube

Sevastopol

Novorossiisk

BULGARIA

Black Sea

N

Istanbul

0   150 km
0   100 miles

Civil War and Foreign Intervention

In the wake of the Revolution, Bolsheviks, Tsarists, Allied troops, anarchists, nationalists and bandits battled for power.

"I am surrounded by moral cowardice, greed and treachery. Some of the Whites are no better than the Bolsheviks."
Alexander Kolchak

Leon Trotsky, the architect of the Red Army. In March 1919 he was appointed Peoples' Commissar for War and Chairman of the Supreme War Council.

Almost as soon as the Bolsheviks seized power in November 1917, former tsarist army officers began assembling their forces in the south and east. Among them was General Kornilov, whose unsuccessful *coup* against the Kerensky government had precipitated the Bolshevik takeover. The commander of the White Russians, as the tsarists were known, Admiral Kolchak, set up his headquarters at Omsk in Siberia while at Rostov, General Kaldin, leader of the Don Cossacks, proclaimed their independence.

Threatened on all sides, the fledgling Bolshevik regime negotiated a peace with the Central Powers. The Treaty of Brest-Litovsk, concluded in February 1918, stripped Russia of most of its western territories including the Ukraine, the Baltic States and Finland. Although the terms were humiliating, they gave the Bolsheviks a breathing space. In March, Kornilov's Volunteer Army captured Yekaterinodar from the Bolsheviks. A few days later he was killed by a shell, and the White forces were driven back, but his successor, Denikin, continued to maintain a powerful presence in the south, assisted by the French. The British and Americans, meanwhile, landed at Archangel and Murmansk, ostensibly to guard supplies stockpiled there before the Revolution.

Following the Armistice in November 1917, the Treaty of Brest-Litovsk was cancelled. Although the Bolsheviks were to recognize the independence of Finland and the Baltic States, they did reoccupy much of the lost territory.

In the Ukraine, however, a nationalist regime, the Rada, had been set up under German protection. The fighting that followed was to be some of the most brutal—and confusing—of the Civil War. It was not merely a straight conflict between Whites and Bolsheviks; Ukrainian nationalists, Cossacks and Nestor Makhno's anarchists all joined the fray, constantly shifting allegiances to meet specific tactical and ideological objectives.

The Czechoslovak Corps, formed to fight alongside the Allies during World War I, found themselves marooned inside Bolshevik territory, reaching Orel—less than 250 miles from Moscow—before being driven back. Yudenich, meanwhile, had launched an attack on Petrograd; that too proved unsuccessful. These offensives were the Whites' last gasp, however. By the end of 1919, the French had pulled out after quarreling with Denikin, and Kolchak had been captured and shot by the Bolsheviks. The Civil War spluttered on, especially in Mongolia and Turkestan, until 1921; but concerted anti-Bolshevik resistance had given way to opportunistic raiding by local warlords.

As the western powers withdrew their forces, all hopes of a lasting victory for the Whites were extinguished. Many thousands from all levels of society fled their homelands, fearful of Bolshevik reprisals against any who might be identified with the old tsarist regime. Many settled in Germany and the other western European states, while others—including Alexander Kerensky—moved on to the United States. The exodus was to continue until 1926 when all emigration was finally banned.

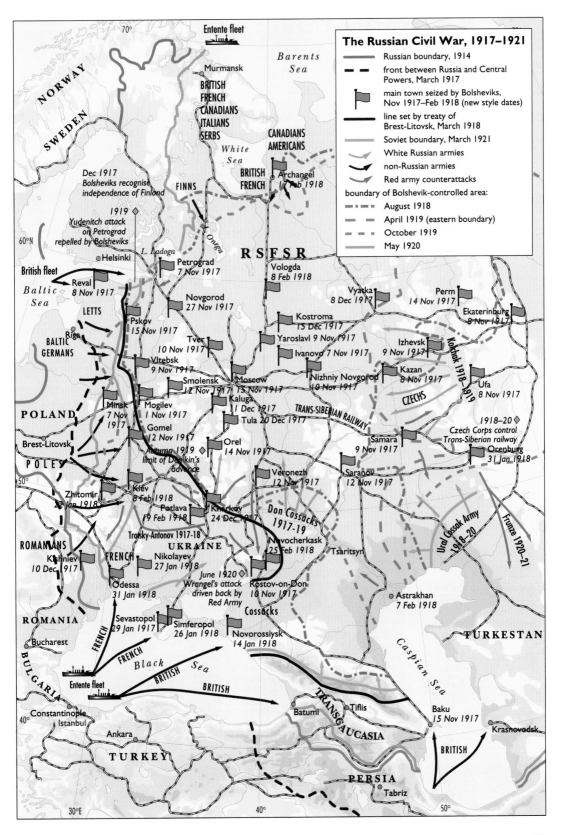

The Russian Civil War, 1917–1921

- Russian boundary, 1914
- front between Russia and Central Powers, March 1917
- main town seized by Bolsheviks, Nov 1917–Feb 1918 (new style dates)
- line set by treaty of Brest-Litovsk, March 1918
- Soviet boundary, March 1921
- White Russian armies
- non-Russian armies
- Red army counterattacks

boundary of Bolshevik-controlled area:
- August 1918
- April 1919 (eastern boundary)
- October 1919
- May 1920

Entente fleet

Barents Sea

NORWAY

Murmansk

BRITISH
FRENCH
CANADIANS
ITALIANS
SERBS

CANADIANS
AMERICANS

White Sea

SWEDEN

BRITISH
FRENCH

Archangel
17 Feb 1918

Dec 1917
Bolsheviks recognise
independence of Finland

FINNS

1919
Yudenitch attack
on Petrograd
repelled by Bolsheviks

60°N

L. Ladoga

Onega

RSFSR

British fleet

Helsinki

Petrograd
7 Nov 1917

Vologda
8 Feb 1918

Vyatka
8 Dec 1917

Perm
14 Nov 1917

Reval
8 Nov 1917

Baltic
Sea

LETTS

Novgorod
27 Nov 1917

Ekaterinburg
8 Nov 1917

Riga
Pskov
15 Nov 1917

Kostroma
15 Dec 1917

Izhevsk
9 Nov 1917

BALTIC
GERMANS

Tver
10 Nov 1917

Yaroslavl 9 Nov 1917

Ivanovo 7 Nov 1917

Kazan
8 Nov 1917

Vitebsk
9 Nov 1917

Smolensk
12 Nov 1917

Moscow
15 Nov 1917

Nizhniy Novgorod
10 Nov 1917

Ufa
8 Nov 1917

POLAND

Minsk
7 Nov 1917

Mogilev
1 Nov 1917

Kaluga
1 Dec 1917

TRANS-SIBERIAN RAILWAY

CZECHS

Kolchak 1918–1919

Brest-Litovsk

Gomel
12 Nov 1917

Tula 20 Dec 1917

1918–20
Czech Corps control
Trans-Siberian railway

POLES

Autumn 1919
limit of Denikin's
advance

Orel
14 Nov 1917

Samara
9 Nov 1917

Orenburg
31 Jan 1918

50°

Zhitomir
22 Jan 1918

Kiev
8 Feb 1918

Veronezh
12 Nov 1917

Saranov
12 Nov 1917

ROMANIANS

FRENCH

Poltava
19 Feb 1918

Kharkov
24 Dec 1917

Don Cossacks
1917–19

Ural Cossak Army
1918–20

Frunze 1920–21

Kishniev
10 Dec 1917

Trotsky-Antonov 1917-18

UKRAINE

Nikolayev
27 Jan 1918

Novocherkask
25 Feb 1918

Tsaritsyn

ROMANIA

Odessa
31 Jan 1918

June 1920
Wrangel's attack
driven back by
Red Army

Rostov-on-Don
10 Nov 1917

Astrakhan
7 Feb 1918

Bucharest

Cossacks

TURKESTAN

BULGARIA

Sevastopol
29 Jan 1917

Simferopol
26 Jan 1918

Novorossiysk
14 Jan 1918

Caspian Sea

FRENCH

FRENCH

Black Sea

BRITISH

Entente fleet

BRITISH

Constantinople
Istanbul

Ankara

Batumi

Tiflis

TRANSCAUCASIA

Baku
15 Nov 1917

Krasnovodsk

40°

TURKEY

BRITISH

PERSIA

Tabriz

30°E

40°

50°

New Economic Policy (NEP)

"The development of international revolution which we predicted is progressing. But this advancing movement is not as direct as we had expected ... "
Vladimir Ilyich Lenin

F*ollowing the tumult of Civil War, War Communism and famine, Lenin introduced economic recovery by "unorthodox" means.*

By 1921 Soviet production had slumped to about one-seventh of pre-war levels. Due to the ravages of war factories had closed and the transport system had broken down. At a time of widespread famine, especially in Ukraine, the southeast and east where millions died—many townspeople fled to the countryside. An emergency relief committee, which included non-Communists, was set up while aid from the American Relief Administration saved numerous lives. The economy had been ruined by Revolution, war, Civil War and the concomitant effects of War Communism, whereby the state assumed control of virtually every branch of national life. Private trade was abolished, and the government directed labour to wherever it was most needed.

In March 1921 Lenin introduced the New Economic Policy (NEP) as a compromise and a modification of communist economic doctrine and practice . NEP implied a mixed economy with a limited system of private enterprise, whereby the state virtually monopolized industrial life, leaving agriculture largely in private hands. This conciliated the peasant majority. Requisitioning was replaced by a food tax and, once this had been satisfied, peasants were left to dispose of their surpluses as they wished. Foreign investment was encouraged and a new monetary unit was introduced in an attempt to stabilize the economy. Small-scale private manufacture was allowed to develop; petty craftsmen and artisans could carry on their own trade legally, leaving the state to control heavy industry, banking and foreign trade. The NEP brought substantial recovery, and by 1926 industry and agriculture had almost reached the levels of 1913.

Yet Lenin would not live to see this marked recovery. In May 1922, Lenin suffered a stroke; a second stroke in December permanently paralysed an arm and a leg. Lenin was incapacitated by a third stroke in March 1923 which ended his political career and he died after a final stroke on 21 January 1924.

A family, reduced to skeletons and clearly at the point of death, begs for alms. Many peasants reacted against the Bolshevik requisitioning of food and planted fewer crops. By 1921 food output had been reduced to less than half the 1913 level and millions faced starvation.

SWEDEN

Barents Sea

Murmansk

0   400 km

0   250 miles

N

White Sea

Archangel

FINLAND

The famine of 1921 and NEP

Soviet Union, 1921

principal famine area

famine conditions widespread

port used by American Relief Administration

route of foreign relief

Industry:

coalfields   oilfields

iron ore   shipbuilding

engineering, armaments and metal industry

areas of industrial concentration

Lake Onega

Kotlas

Gulf of Bothnia

60°N   Helsinki
Hango
Gulf of Finland
Reval

ESTONIA   *Lake Peipus*

Lake Ladoga

Petrograd
St Petersburg

Vologda

Viatka

Perm

Ekaterinburg

60°

Lake Ilmen

Kostroma

Riga
LATVIA
W. Dvina

Vitebsk

Gorky

Kazan

Ufa

Chelyabinsk

LITH.

Nieman

POLAND

Mogilev

Moscow

Russia

Simbirsk

Samara

Belorussia

Tula

Penza

Orenburg

Chernigov

U   S   S   R

Veronezh

Saratov

50°

Kiev

U k r a i n e

Dnieper

Poltava

Kharkov

Don
Don Cossacks

Izium

Volga

Tsaritsyn

Guryev

Rostov

ROMANIA   Odessa

Kuban Cossacks

Astrakhan

Sea of Azov

Kuma

Sevastopol

Novorossiisk

C a s p i a n   S e a

Turkestan

B l a c k   S e a

Ural

Tiflis
Tbilisi

Baku

Krasnovodsk

Batumi

Transcaucasia

40°

Ankara

T U R K E Y

105

Creation of USSR and Rise of Stalin

As Lenin's health waned, Stalin rose to power like a Machiavellian puppet master, pulling the strings.

"He has concentrated too much power in his hands... I am not sure that he always knows how to use power with sufficient caution."
Lenin, speaking of Stalin in 1923

Late in 1917 power was transferred to the second All-Russia Congress of Soviets which elected a new government, the Council of Peoples' Commissars (or *Sovnarkom*), led by Lenin. Early in 1918 the third All-Russia Congress of Soviets issued a Declaration of Rights of the Toiling and Exploited Masses which proclaimed Russia a republic of soviets of workers, soldiers and peasants' deputies. In July the fifth congress adopted a constitution creating the Russian Soviet Federated Socialist Republic (RSFSR). Local soviets elected delegates to a provincial congress of soviets, each of which elected the Executive Committee which acted in the intervals between sessions of congress and the *Sovnarkom*.

Due to the exigencies of war the state had been obliged to relinquish Finland, Latvia, Estonia and Lithuania, and the Polish territories which had gained independence. Western Ukraine and western Belarus were ceded to Poland, Bessarabia to Romania, and the Kars-Adakhan area and Transcaucasia to Turkey.

In the course of the Civil War, other soviet republics were set up in eastern Ukraine, eastern Belarus and Transcaucasia, which first entered into treaty relations with the RSFSR and then, in 1922, joined with it in a Union of Soviet Socialist Republics (USSR). Later in the 1920s, three Central Asian republics became Union republics.

In accordance with the Constitution (1924) and the Union treaty (December 1922), the larger non-Russian nationalities were offered equality of status within the Soviet Federation, with the right to secede and considerable administrative and cultural autonomy. By becoming part of a socialist federation, the nationalities would lose their state sovereignty. A number of Bolshevik policies promoted national language, education and culture, encouraging the development of indigenous personnel rather than promoting Russian or Russified "cultural pluralism", and thus obviating the danger of "Great Russian chauvinism" in the non-Russian republics. By the late 1920s this was viewed by Stalin as a threat to his policies.

Stalin had served as Peoples' Commissar for the Nationalities before his appointment as General Secretary of the Party at the XIth Party Congress in April 1922. During his illness (May 1922-January 1924), Lenin became aware of brutalities carried out in Georgia in 1921 by Ordzhonikidze, one of Stalin's men, and accused Stalin of Russian chauvinism in his "testament", although this was not revealed until Khrushchev's Secret Speech at the XXth Party Congress in February 1956.

As Lenin faded into virtual imbecility, Stalin carefully fostered the image of a warm, modest and sincere man committed to the good of the Soviet people. His success in maintaining this facade seems remarkable in light of the fact that his ambition and paranoia resulted in the death of millions.

Following Lenin's death, Stalin seized the main chance by consolidating his power base as General Secretary of the Party. First he opposed Trotsky, the key opponent for the leadership of the Communist Party, by allying with Zinoviev and Kamenev. Discredited by the Triumvirate in 1925, Trotsky was forced to relinquish his post as Commissar for War, and two years later was expelled from the Communist Party along with 80 of his followers. (Exiled to Alma Ata, then from the USSR, Trotsky was eventually killed in 1940 by a Stalinist assassin.) Allying with Bukharin, Stalin turned on Zinoviev and Kamenev. By 1929 Stalin had gained total power within the party.

The climax of the first of Stalin's carefully orchestrated show trials, as the presiding judge, Andrei Vyslinsky reads out the death sentence to the eleven defendants.

Overcoming national resistance

declared independence from Russia 1917

declared independence from Russia 1918

briefly independent, then incorporated into Soviet Union

land lost by Russia after WWI

✳ centre of anti-Bolshevik revolt

strategic rail link to east and south

Bashkiria

Soviet People's Republics, established 1917

Kazakh nomads flee Tsarist conscription and later Bolshevik rule

area of anti-Soviet revolt 1923–1931

◇ *1917 nationalist uprising 1918 alliance with anti-Bolsheviks 1919 accepts Bolshevik assurances of autonomy 1920 Bolsheviks suppress all resistance*

0   800 km
0   500 miles

N

◆ *Nov 1917 controlled by Bolsheviks*

1917 ◇ *Bokhara and Khiva established as Soviet Peoples' Republics*

◆ *1923–31 area of anti-Soviet revolt*

NORWAY
SWEDEN
FINLAND
NETH.
DENMARK
Oslo
Copenhagen
Stockholm
Gulf of Bothnia
Baltic Sea
Berlin
GERMANY
Helsinki
Danzig
Barents Sea
White Sea
ARCTIC CIRCLE
Tallinn
ESTONIA
East Prussia
Kaunas
Riga
LATVIA
LITHUANIA
Lake Ladoga
Lake Onega
Warsaw
Petrograd
Leningrad
POLAND
W Dvina
U S S R
Pripat
Smolensk
ROMANIA
Bessarabia
Kiev
Moscow
Volga
Odessa
Dnieper
Sevastopol
Rostov
Saratov
Ufa
Don
Ural
Chelyabinsk
Irtysh
Trans-Siberian Railway
Orenburg
Omsk
Novosibirsk
Black Sea
Caucasus
Astrakhan
Aktyubinsk
Barnaul
GEORGIA
Tbilisi
TURKEY
Caspian Sea
Karaganda
ARMENIA
Yerevan
AZERBAIJAN
Baku
Kazalinsk
Semipalatinsk
Aral Sea
Syr Darya
Lake Balkhash
SYRIA
Krasnovodsk
Khiva
Perovsk
Baghdad
Tehran
Amu Darya
Tashkent
Verny
Tien Shan
Issyk Kul
Pishpek
Ürümqi
IRAQ
PERSIA
Ashkhabad
Mery
Bukhara
Skobelev
Samarkand
Kokand
CHINA
Kushk
Diushambe
Sinkiang
AFGHANISTAN
BRIT. INDIA

Industrialization and Collectivization

"We are 50 or a 100 years behind the advanced countries. We must make up this leeway in ten years. Either we do it or they crush us."
Joseph Stalin

The Stalinist organization of the Soviet economy was realised through a series of Five Year Plans involving the expansion of industry and the collectivization of agriculture.

Collectivization was forced upon the countryside; peasants were dispossessed and collectives of state farms were established. In industry priority was given to producers' goods over consumer goods and all available resources were channelled into heavy industry. Although production rates generally increased, this was often at the expense of quality.

The plans were ambitious, resulting in social revolution and considerable hardship for workers. Yet the 1930s made for an heroic age in which workers and peasants accused of wrecking or industrial sabotage were either exe-

A Soviet industrialization poster of the early 1930s. Although industrial output was increased under Stalin, the emphasis on some sectors at the expense of others meant that progress was neither balanced nor consistent.

Industries:

- coal mining
- oil field
- textiles
- chemicals
- non-ferrous metals
- beet-sugar processing
- iron ore mining
- metal processing
- machine building
- electric power station

A party of female labourers return from work on a collective farm.

Collectivization

▬▬ boundary 1939	••••• northern limit of agriculture
▨ principal agricultural areas	══ important rail link
▨ main area of collectivization	▬▬ northern sea route
	● concentration of industry

cuted or sent to the Gulag, whilst those who reached outstanding levels of production, the "Stakhanovites", were rewarded with pay incentives and better housing conditions.

By March 1930, 55 per cent of peasants had been collectivized, although in some areas, such as Kazakhstan, Uzbekistan and the Caucasus the percentage was much lower. By 1934, about three-quarters of the peasants' farms had been brought into collectives. Yet most peasants resented collectivization and the traditional divide between peasant and urban mentalities widened as Soviet officials from the towns coerced peasants into joining collectives and began to requisition their livestock and tools.

Collectivization changed the lives of over 75 per cent of the Soviet population. By 1937 almost all cultivated land was in collective farms (*kolkhozes*) or the state farms (*sovkhozes*) which were more directly controlled on factory lines by the state. Many peasants did as little as possible in the work brigades of the collective farms, and put everything into tending their own plots. In 1938, although the private plots of land made up only three per cent of the area farmed, they contained over half the cattle.

In the Soviet Union many collective farms were inefficient. On the eve of the German invasion in 1941, the agricultural production still had not returned to 1928 levels. The amount of food produced fell sharply. Rather than hand over their beasts to the collectives, peasants slaughtered them. Famine followed in 1932 and 1933, yet despite the shortages, the government insisted on the peasants supplying as much grain as before to feed the towns and continued to export grain for the industrialization drive.

The First Five-Year Plan was presented to the Party in 1928. It laid down the targets for the production of everything considered necessary for an advanced economy. Traditional industrial areas such as Leningrad, Moscow and the Donbass expanded, whilst technological and engineering projects were taken to the remote areas of Kazakhstan and the Caucasus. New industrial centres were developed in the Urals, Kuzbass and the Volga. The output of electricity almost trebled between 1928 and 1932, and within ten years this rate of output had increased sixfold.

The Second Five Year Plan (1933–37) got off to a difficult start because of the 1932/33 famine; the result of poor yields and requisitioning from a peasantry lacking incentives. Concomitant with the general chaos of collectivization were a transport crisis and severe shortages in many industries. Again, priority was given to heavy industry, although more consumer goods were produced than in the First Five Year Plan. The defence industry achieved most, trebling its output between 1933 and 1937, although this resulted in other sectors of the economy becoming depressed since the best scientists, engineers and workers were recruited to the defence sector.

The Third Five Year Plan was formally adopted at the 18th Party Congress in 1939, but was cut short by the German invasion which began in June 1941. Nevertheless, on the outbreak of war, the foundations of an industrial base had been well laid.

Foreign Policy

In the face of western appeasement of aggressors and hostility to Soviet Russia, Stalin moved towards an alliance with Nazi Germany.

"What they want is a treaty in which the USSR would play the part of a hired labourer bearing the brunt of the obligations on his shoulders."
Zhdanov, *Pravda*, June 1938

In the early 1920s Soviet Russia entered a period of diplomatic isolation following the publication of secret wartime treaties, the takeover of foreign-owned industries and the repudiation of foreign debts incurred during the tsarist period. Allied intervention in the Russian Civil War and French support for Poland during the Russo–Polish war caused further hostility. Throughout the 1920s Soviet relations with the British, French and American governments remained cold.

Nevertheless a period of good relations developed between Germany and Soviet Russia, following the Treaty of Rapallo on 16 April 1922. These relations were strengthened by a pledge of economic co-operation and the renunciation of financial claims on each side. Mutual dislike of the Polish Republic engendered military co-operation, the Red Army and Black *Reichswehr* taking part in joint manoeuvres, and enabling Germany to circumvent the Versailles Treaty by making armaments and training its tank and air-crews, in secret, on Soviet territory.

Following Japan's invasion of Manchuria in 1931, the main threat to Soviet security came from the east. But Moscow's fear that an agreement between Japan and Chiang Kai-shek might result in a Japanese invasion of Siberia or Outer Mongolia was ended in July 1937, when Japan attacked Chiang's *Guomindang*. The USSR signed a non-aggression pact with China and left them to pursue their war with Japan.

Meanwhile, with the rise of National Socialism in Germany, Moscow negotiated non-aggression pacts with France, Poland, Finland and Estonia in 1932 and Italy the following year. These added to earlier agreements with Lithuania and Finland. The creation of a popular front to check the advance of fascism followed shortly after the "Night of the Long Knives" (30 June 1934), when Hitler secured his position and gained the support of the German officer corps.

In 1934, the Soviet Union improved its international situation by joining the League of Nations and in May 1935 signed the Franco–Soviet treaty of mutual assistance, which included a commitment to aid Czechoslovakia, though France subsequently released the USSR from its obligation. In 1936 the Spanish Civil War broke out and the Soviet Union alone went to the aid of the Republicans. In the meantime Germany, Italy and Japan signed an Anti-Comintern pact. Everywhere the

A Republican poster of the Spanish Civil War acknowledges the support of the Soviet government in its fight against the Fascist General Franco. In fact, Soviet aid was strictly limited, consisting almost solely of small shipments of equipment. In contrast, Franco's Nationalists received active assistance from both Mussolini's Italy and Hitler's Germany.

North Sea

UNITED KINGDOM

London

HOLLAND

BELGIUM

LU

Paris *Seine*

FRANCE

S

Rhône

San ebastian

Ebro

SPAIN
Madrid

Barcelona

Mediterrane

0°

fascists seemed to have gained the initiative.

Having successfully annexed the Czech Sudetenland in the face of Anglo-French timidity in September 1938, Hitler focussed on the incorporation of Danzig's German population into the Reich. In response to Poland's appeal for help, France and Britain guaranteed its frontiers, although neither country could achieve much without Soviet support. Stalin was now offered the choice of either an agreement with France and Britain, or one with Germany. Suspicious that the western powers might ultimately abandon the Soviet Union to face German aggression alone, Stalin chose to sign a non-aggression pact with Germany. The treaty contained a secret protocol placing Estonia, Latvia, Finland and the eastern part of Poland within the Soviet sphere of influence, and Lithuania within the German sphere.

Foreign Policy, 1920 s–1930s

⟋⟋⟋	country that signed Soviet trade agreement during 1921
△	centre of German military training
▨	signed Non-Aggression Pact with Soviet Union in 1932
▨	signed military assistance treaty with Soviet Union in 1935
⟋	route of Soviet aid to Republicans in Spain
▨	allied to Soviet Union in Aug 1939 by Nazi-Soviet Non-Aggression Pact

Stalinist Repression

In an atmosphere of fear and repression Stalin set about the total annihilation of all possible opposition to his leadership.

"Through the sewer pipes the flow pulsed. Sometimes the pressure was higher than had been projected, sometimes lower. But the prison sewers were never empty."
Alexander Solzhenitsyn, *The Gulag Archipelago*

The Great Purges or *Yezhovshchina*, named after the eponymous Commissar for Internal Affairs, was a series of show trials and executions that took place in the second half of the 1930s. The terror was unleashed in December 1934, following the murder of Kirov, a close associate of Stalin, and chief of the Leningrad Party organization. Recent research has disproved the long held belief that Stalin himself had ordered Kirov's death. During the reign of terror that followed, the Soviet security organizations and prison system were mobilized in an attempt to annihilate all opposition and dissidence. Those whom Stalin considered as personal rivals were liquidated and millions of Soviet citizens were sent to the Gulag Archipelago, a network of prison camps stretching across the north of European Russia and into north-eastern Siberia, where they suffered long periods of imprisonment. Camp inmates were often cruelly treated, working up to 16 hours a day and fed according to what they produced.

All the members of Lenin's politburo, apart from Stalin and Trotsky, were executed. This included some of the most prominent Soviet personalities, such as Bukharin, Kamenev, Rykov and Zinoviev, along with most of the surviving Old Bolsheviks and many members of the upper and middle levels of

In the slave labour camps prisoners were made to work on projects designed to benefit the Soviet nation. One such project, the White Sea Canal, cost thousands of lives but proved to be too shallow for the ships of the Baltic Fleet. Prisoners were forced to labour in appalling conditions, denied adequate food or clothing. Those unable to work were executed in mass shootings.

1 Crimean Tartar Autonomous Republic
2 Karachai Autonomous Republic
3 Meskatians
4 Chechen-Ingush Autonomous Republic

the Communist Party. Likewise more than half the generals of the Red Army—including Tukhachevshy, Chief of the Red Army and Admiral Orlov, Commander-in-Chief of the Red Navy—were shot.

The Red Army was severely damaged by the Purges. The loss of the officer corps left it poorly organized, badly commanded and demoralized. The Winter War of 1939–40 revealed to all the incompetence of the Red Army. Khrushchev later recounted how Voroshilov, People's Commissar for Defence, in a rage blamed Stalin for the Red Army's setbacks and heavy casualties, shouting: "You're the one who annihilated the Old Guard of the army; you had our best generals killed!"

The system of repression and the spread of the Gulag system of "corrective labour" camps represented the bloodiest period of the Stalinist regime and the high point of the growing totalitarian nature of Soviet society. It has been estimated that about 10,000,000 people were arrested and 3,000,000 executed before the German invasion of the Soviet Union in June 1941.

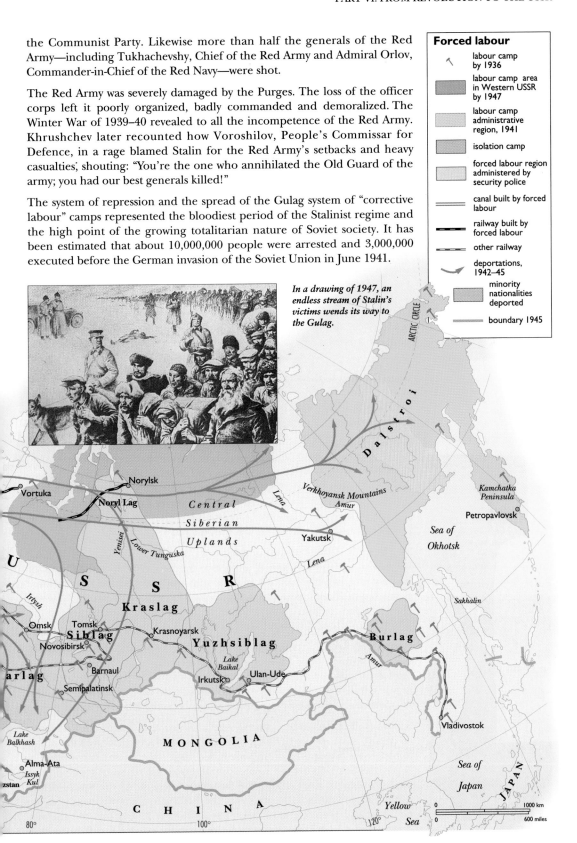

Forced labour

- labour camp by 1936
- labour camp area in Western USSR by 1947
- labour camp administrative region, 1941
- isolation camp
- forced labour region administered by security police
- canal built by forced labour
- railway built by forced labour
- other railway
- deportations, 1942–45
- minority nationalities deported
- boundary 1945

In a drawing of 1947, an endless stream of Stalin's victims wends its way to the Gulag.

113

VII: War, Peace and Dissolution

"Our detractors used to say that the only reason we were able to defeat Paulus' colossal army at Stalingrad was that we had the Russian winter on our side ... Ever since Russia turned back Napoloen's invasion, people claimed that winter was our main ally."

Nikita Khrushchev

Execution of Russian 'partisans' by members of an Einsatzgruppe shortly after the commencement of Operation Barbarossa. The intense hatred which existed between the German invaders and the Soviet defenders led to atrocities being committed on both sides and capture meant almost certain death whether through deliberate policy or simple neglect.

In 1917 the rule of the tsars finally ended in Russia, but their autocracy was replaced by an even more repressive regime.

Shortly after Germany had launched Operation *Barbarossa* in June 1941, millions of Soviet citizens and great quantities of machinery were transferred beyond the Urals, to the Caucasus and Central Asia where new factories were established. Eventually these would produce arms and munitions, especially the new T-34 tank which proved far superior to its German equivalents.

The Germans, influenced by National Socialist theories of racial superiority, put into practice by their *Einsatzgruppen* (Task Force), treated the eastern Slavs as *Untermenschen* (sub-humans). They failed to win hearts and minds in the areas they occupied. Only a handful of Ukrainians, Russians and Belorussians would join the German cause, either out of hatred of Stalinism or in order to survive the terrible conditions endured as prisoners in German concentration camps. In response to German cruelty, partisans operated behind the front wreaking havoc on German lines of communication.

The *Wehrmacht* suffered its first setback during the harsh winter of 1941 when German soldiers in their summer uniforms froze to death on the outskirts of Moscow. They would be pushed back between 80 and 300 kms. In the meantime, Leningrad endured the first winter of its dreadful two-and-a-half year siege.

A change in Stalin's own policies came out of the need for Soviet unity. On 3 July 1941, in his first radio address to the nation in three years, Stalin had appealed directly to Russian patriotism and religion. For the time being persecution ceased, and a call was made for the defence of the motherland. To restore morale and discipline in the army, the old rank structure and the saluting of superiors was restored. To ensure that food supplies were distributed more equitably, rationing was introduced.

The Germans momentarily regained the initiative in the first half of 1942, advancing south in an attempt to capture the rich oilfields of the Caucasus. The Soviets, however, fought back courageously, supported now by large quantities of Allied military equipment and food brought into the country either via Iran or the Pacific, or after treacherous maritime voyages to the White Sea ports of Archangel and Murmansk.

Despite the initial unpreparedness of the Red Army, and the concomitant suffering with tremendous loss of life and territory in European Russia, the tide of war eventually turned with the dawn of 1943. After the great victory of Stalingrad, followed five months later by the decisive tank battle at Kursk, the Germans would never regain the initiative on the Eastern Front.

Stalin asked the Allies to open up a Second Front to relieve the Red Army of the full brunt of the fighting, but this did not materialize until the Allied landings in North Africa in November 1942, and in southern Italy in September 1943. In the meantime, the victory at Stalingrad had served as an encouragement, not only to the Soviet people, but to the whole Allied cause. By the end of February 1943, the Red Army regained nearly all the land that had been lost to the German advance of 1942. From the summer onwards, the Red Army maintained its superiority over the *Wehrmacht* in both men and equipment as it drove relentlessly on to Berlin.

The Soviet theatre of operations remained confined to Europe until 8 August 1945, when Moscow declared war on Japan and Soviet troops invaded Japanese-held Manchuria. The war in Asia came to an end six days later, following the bombing of Hiroshima and Nagasaki.

The contribution of the Soviet Union to the final Allied victory was enormous. At an estimated loss of more than 20 million Soviet citizens, no country had suffered more. The war had caused immense material losses. In an attempt to repair the ravages of the war and to prepare for the future, in 1946 Stalin launched a plan of National Reconstruction.

Red Army soldiers raise the "hammer and sickle" above the ruins of the German Reichstag *in Berlin, May 1945. The Soviet offensive to take Berlin began on 16 April and ended on 2 May and cost an estimated 100,000 Soviet lives.*

The fourth Five-Year Plan (1946–50) concerned post-war reconstruction, while at the same time emphasizing the production of military equipment in order to maintain the strength of the armed forces. The areas of heavy industry that had been established in central and eastern Russia—and which had proved vital to the Soviet war effort—continued to be developed,

whilst equipment was requisitioned from Germany and the occupied areas of eastern and central Europe. By 1948 Soviet factories were producing as much as they had in 1940; by 1952 industrial production had doubled the pre-war total. Despite this achievement, there were still shortages, especially in the areas of housing and consumer goods.

The post-war period was also a time in which Stalin's cult of personality reached its apogee. This was largely based upon exaggerated reports of his wartime leadership, and to the detriment of his military commanders such as Marshal Zhukov whom he saw as a potential rival. Hopes for a relaxation in the authoritarian regime were dashed by a return to the repressive security methods of the 1930s. Critics of Stalin and of his government were sent to the Gulag, along with those Soviet soldiers who had spent time in German prison camps, one of the most famous examples being former artillery officer, Alexander Solzhenitsyn. The climate of fear and suspicion grew in 1948 with a new wave of arrests. Nobody was above suspicion. The wife of Foreign Minister Molotov was sent into exile and Stalin even had his daughter Svetlana's lover sent to an Arctic labour camp.

At the beginning of 1953 there were fears that Stalin was preparing another purge, which would involve leading Communists implicated in the so-called Doctors' Plot, whereby Kremlin doctors—many of whom were Jewish—were alleged to have planned to kill the Soviet leadership. With Stalin's death on 5 March 1953, the allegations were dropped and the doctors rehabilitated within a month.

There followed a period of collective leadership between Malenkov, Molotov, Kaganovich and Khrushchev. Each developed his own policies in an attempt to assure his own position. There were disputes over how to run the economy and how to raise living standards. For example, Malenkov felt it was time to favour the development of light industry, while Khrushchev wanted to raise agricultural output. Almost immediately the NKVD was brought under Party control; its chief, Beria was arrested and executed. Above all there was a wish to break away from the stranglehold of Stalinism. In February 1956, Khrushchev in his Secret Speech to the XXth Party Congress questioned many of Stalin's policies, denounced his crimes and the cult of personality, and yet avoided any condemnation of the system, so that no criticism was made of the period before 1934. The contents of the Secret Speech leaked out and led to a period of confusion in the Soviet

Nikita Khrushchev and aides examine the maize crop on the Novi Zhylta collective farm in 1963. As in a number of other projects, Khrushchev's enthusiasm masked an economic ineptitude. The emphasis on maize pushed up the prices of meat and dairy products and crowds demonstrating against the maize policies were fired upon.

sphere of interest in eastern and central Europe, with a wave of popular unrest in Poland, East Germany and Hungary.

By 1957 Khrushchev assumed power by dismissing Malenkov, Molotov and Kaganovich from the Politburo, becoming Chairman of the Council of Ministers in 1958. Khrushchev's rise to power had been far less damaging than that of Stalin.

There followed a period of de-Stalinization: the "Russian thaw" of the 1950s, when attempts were made to free the Soviet Union from its Stalinist past. Later, in 1961, Stalin's body was removed from the mausoleum he shared with Lenin and buried secretly in the Kremlin wall. In keeping with the spirit of the times, Stalingrad was renamed Volgograd. Khrushchev had worked closely with Stalin in the past, but he wanted to do away with terror and to be seen as a benign leader. Some legality was introduced into police procedures and many prison camps closed, although coercion, arrest, imprisonment and surveillance remained a part of the system.

Nevertheless, for the ordinary Soviet citizen the 20 years from 1954 to 1974 was a period of rising living standards, optimism and peace. During the Khrushchev period the official belief, expressed at the XXII Party Congress of 1961, was that the Soviet Union would overtake the United States in per capita production by 1980, and that Communism could be achieved within a period of 20 years. The later 1950s was a time of achievement and Soviet pride; the Soviets were the first into space and Soviet technological advancement came as an immense shock to the United States.

In the meantime, special attention was given to agriculture, an area in which Khrushchev had considerable experience. There followed a programme of partial decentralization, with the introduction of widespread changes in regional economic administration and the amalgamation of collective forms (*kolkhozy*), officially to facilitate administration and to improve efficiency, in keeping with Bolshevik ideology that large was best. The "Virgin Lands" programme was launched in a search for popular and speedy economic results and an attempt to solve problems of food supply.

Khrushchev may have been skilled in agricultural matters, but he had little or no experience in foreign policy. Despite this, Soviet influence in the world increased under Khrushchev, although he brought the world to the brink of war.

When Stalin died, the Cold War thawed slightly. At the Geneva summit in 1955 agreement was reached to establish Austria as a neutral state. The granting of Austrian independence, her neutrality and the influence of "de-Stalinization" in turn created an atmosphere of hope and nationalism in Hungary. Khrushchev confronted the Hungarian Revolution in 1956 with force and determination. Soviet tanks were sent in and the streets of Budapest shelled by the Red Army. Although 190,000 Hungarians fled into exile and their leader, Imre Nagy, was executed, Hungary was rapidly drawn back into the Soviet fold.

In 1962 the Soviet Union came perilously close to war with the United States over the Cuban missile crisis when the USSR planned to place medium-range rockets on Cuba. Eventually, in the face of President John F. Kennedy's determined opposition, Khrushchev backed down and withdrew the missiles. The fact that he had taken decisions without consulting his military advisers, together with his submission to Kennedy, severely damaged his reputation among Party colleagues at home.

"What can perestroika possibly mean down here? We long since ceased to believe in the system."
from Susan Richards, *Epics of Everyday Life,* 1990

Despite this, Khrushchev unwisely decided to travel abroad for a period of 135 days, leaving his many opponents plenty of time in which to prepare a coup for which they felt they had considerable justification. Apart from the Cuban fiasco, Krushchev's colleagues were also critical of his failure to raise agricultural production as much as hoped. The final straw was the humiliating need to import grain from the West.

In October 1964 Khrushchev was dismissed from office by the Presidium, becoming thereafter practically an "un-person". A period of collective leadership followed, shared between Leonid Brezhnev, Aleksei Kosygin and Nikolai Podgorny. The Party entered a period of maintaining the status quo based on "sensible policies", rather than campaigning for a future society. After so much upheaval, with rapid industrialization, repressive collectivization, the horrors of the Patriotic War, the difficult post-war years, the continuous agricultural and industrial reforms and the so-called "hare-brained" schemes of Khrushchev, the time was ripe for stability, certainty and—above all—a slowing down of innovative policies.

Eventually, Leonid Brezhnev came to power as First Secretary of the Party. Under his leadership there was far less reform in domestic policies; throughout the 1970s the living standard continued to rise, although the technology gap by the end of the 1970s was growing.

The Soviet Union expanded its influence in the Third World, especially in Angola, Mozambique, Vietnam and Afghanistan. Unlike Khrushchev, Brezhnev avoided confrontations with the military, whom he employed as foreign policy advisers. In the military sector the Soviet Union had successfully achieved nuclear parity with the United States by the late 1960s. This parity, however, led to a huge drain on resources. The Soviet Union's GNP was only half the size of that of the United States, yet it had to devote a far greater proportion of its resources to defence in order to compete. A two-track economy developed between military and consumer needs. The war in Afghanistan proved a costly failure. Maintaining Soviet control over the economically weak eastern European buffer zone was costing too much and produced little of any benefit. Star Wars eventually constituted an even greater strain on the economy.

The period 1963–78 witnessed a gradual détente, or relaxation of East–West tensions, despite the Soviet invasion of Czechoslovakia in 1968. The Conference on Security and Co-operation in Europe held at Helsinki in August 1975, accepted the need for developing economic relations and recognised the post-war frontiers of Europe. The mood of the conference was one of conciliation and détente. As part of Willy Brandt's *Ostpolitik*, the Federal Republic of Germany accepted the western frontier with Poland, and West Berlin was no longer interpreted as a part of the Federal Republic.

Forces of the Soviet Army patrol the mountain passes of Afghanistan shortly before their final withdrawal. Throughout ten years of costly skirmishing with the Mujahidin, the Soviet government proved unwilling to commit the large forces necessary to subjugate such a country. An ignominous retreat was undertaken in 1988–89.

The East–West relationship started to cool with the Yom Kippur War of October 1973 and Soviet assistance to anti-Western movements in Africa—especially in the former Portuguese colonies of Mozambique and Angola and Soviet involvement in the Ethiopian–Eritrean conflict. But it was the Soviet invasion of Afghanistan in December 1979 which ended the period of détente.

In an attempt to reduce the cost of the strategic arms race, the super powers had embarked on the Strategic Arms Limitation Talks (SALT) in the early 1970s. Although the SALT talks did not reduce East–West tension, they at least regimented and limited the number of missiles, bombers and war-

Mikhail Gorbachev is greeted by Prime Minister Margaret Thatcher on the second day of his state visit to the United Kingdom in April 1989. Not since the days of Khrushchev had a Soviet leader revealed such a willingness to be portrayed as an international diplomatist. Gorbachev's desire to introduce a thaw to the Cold War earned him the Nobel Peace Prize.

heads held by both sides. Nevertheless, they were limited in their scope; the talks did not bring a halt to the arms race, nor did they lead to disarmament. When the agreement was signed by Brezhnev and Carter in 1979, East–West relations were already degenerating.

The problem was an ageing leadership, increasing bureaucratization and the stress placed upon consensus decision-making. The Soviet Union had lost its dynamism of the late 1950s and early 1960s. Mass organizations such as the *Komsomol* (the All-Union Leninist Communist League of Youth), the trade unions and cultural organizations had degenerated into mere extensions of the administrative system and the Party, and failed any longer to inspire. There was growing disillusionment.

Was the Soviet Union becoming moribund? There were social and economic problems but these were not discussed, thus their extent was unknown. Racism, feminism, violence, crime, drug abuse and abortion; issues addressed in the West, were officially non-subjects in the Soviet Union.

The Soviet Union could not keep up with the changes taking place in the world economy in the 1980s, with the transfer from labour to services, from heavy industry to information technology. The war in Afghanistan particularly affected the Soviet economy both as a result of sanctions being imposed on the USSR by the United States, and the drop in sales of technology. For example, when Gorbachev came to power in 1985, there were 50,000 personal computers in the Soviet Union compared with 30 million in the United States.

Brezhnev died in 1982; he was replaced in quick succession by Andropov who was about to open dialogues with the West—and, in February 1984, by the ailing Chernenko. When Mikhail Gorbachev came to power in 1985 he wanted to reform Communism, but not replace it. Like Khrushchev, he wanted to reshape the Soviet Union without destroying it. There followed a revolution from below, rather than one controlled from above. In an attempt to overcome economic stagnation Gorbachev tried to discipline the Soviet people. When this failed he launched the idea of *perestroika*, or "restructuring", but was unable to restructure those in charge as they consistently frustrated his orders. To weaken them, he introduced the policy of *glasnost* leading to democratization and open discussion. Once people were able to express their ideas more freely, they were able to criticize and demand even more change.

The collapse of Communism in response to the new spirit of reform ushered in the collapse of the highly centralized Soviet empire. From December 1986 until Gorbachev surrendered office in 1991, millions took to the streets to express their discontent with the ethnic, economic and political policies of the Soviet Union. Hundreds died in the ensuing inter-ethnic conflicts, or in actions carried out by the Red Army and Soviet police forces in Azerbaijan, Armenia, Georgia and Kazakhstan.

With the failed coup of August 1991 it was clear that the Communist Party had lost its authority and that the Soviet Union was disintegrating. In December 1991, it ceased to exist, giving place to a loose alliance called the Commonwealth of Independent States (CIS) which included the Russian Federation (or Russia), Ukraine and Belarus, but excluded the three Baltic Republics and Georgia.

Barbarossa

With France conquered and Britain seemingly on the verge of collapse, Hitler at last was able to turn his attention eastwards.

"Our troops are fighting heroically against an enemy armed to the teeth with tanks and aircraft. Overcoming numerous difficulties, the Red Army and Red Navy are self-sacrificingly fighting for every inch of Soviet soil."
Joseph Stalin, July 1941

By the summer of 1940 only a beleaguered Britain stood between Germany and total victory in the West, and Hitler felt able to commence operations against the USSR. Operation *Barbarossa*—the German invasion of the Soviet Union—was the largest single campaign of the World War II, involving over three million Axis troops and almost five million Soviets. German forces advanced on three fronts against Leningrad in the north, Smolensk and Moscow in the centre, and Ukraine in the south.

The Russo–German Non-Aggression Pact, signed by Joachim von Ribbentrop and Vyacheslav Molotov in August 1939, gave the Soviet Union nearly two years to increase armaments production and to repair some of the damage caused to the Red Army by the loss of over 40,000 officers during the purges. Yet when the invasion commenced, the 11 Soviet armies close to the border were poorly organised, insufficiently equipped and incompetently led.

Stalin had been warned of imminent invasion but he refused to listen to his advisers, persisting in his belief that rumours of impending conflict with Germany were unfounded and provocative and that relations with Germany should remain normal. As a result the massive opening artillery bombardment in the early hours of Sunday, 22 June 1941, caught the Red Army completely by surprise. Units were soon overwhelmed and whole armies were rapidly out-manoeuvred as soldiers and equipment were captured and destroyed. Meanwhile, the *Luftwaffe* gained air superiority by destroying an estimated 2000 Soviet aircraft within the first 48 hours.

German armies overran Estonia, Latvia and Lithuania, Belorussia and Ukraine, and advanced into the Russian Federation. By 29 June, Minsk had fallen. Pskov was taken on 8 July and on 14 July tanks were within 80 miles of Leningrad. By 16 July Smolensk had fallen to General Guderian. By occupying the European part of the Soviet Union, the Germans deprived the Soviets of their agricultural and mining heartland and the heaviest concentration of industry and the bulk of the railway network.

A Russian propaganda poster of 1942 by V. Koretsky. The slogan reads "Warrior of the Red Army, Save us!".

In 1941 Soviet equipment was inferior in quality to that of the Germans and there were shortages of rifles, machine guns, heavy artillery and transport. Soviet tanks were not as well armed or armoured, and few Soviet planes and tanks were equipped with radios. But there were deeper reasons for the initial Soviet disaster: a desperate shortage of officers and trained personnel and low morale. Few lessons had been learned from the costly Finnish war, and many Soviet commanders had been promoted beyond their abilities to replace those executed in the purges.

But by the autumn the Germans were losing their momentum. Valuable campaigning time had been lost because of the exigencies of invading Yugoslavia and Greece in the spring, and the Germans were not equipped for a winter campaign. Supplies were hampered by the need to convert the Soviet railway system from its broad, non-standard gauge. In the meantime, contact between German Army groups—Centre and South—had been broken by the Pripet Marshes, whilst von Rundstedt's forces in the south faced unexpected resistance from the Soviet Fifth Army.

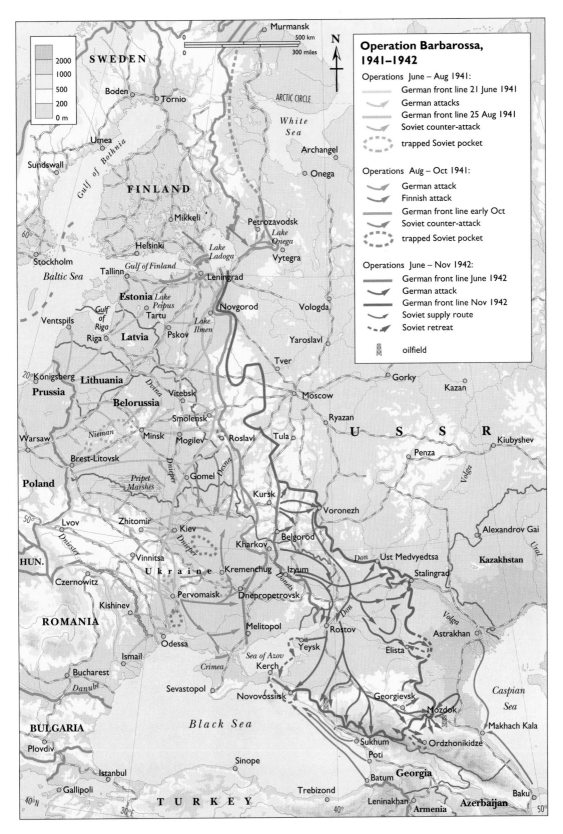

Operation Barbarossa, 1941–1942

Operations June – Aug 1941:
German front line 21 June 1941
German attacks
German front line 25 Aug 1941
Soviet counter-attack
trapped Soviet pocket

Operations Aug – Oct 1941:
German attack
Finnish attack
German front line early Oct
Soviet counter-attack
trapped Soviet pocket

Operations June – Nov 1942:
German front line June 1942
German attack
German front line Nov 1942
Soviet supply route
Soviet retreat

oilfield

Stalingrad to Berlin

Fought in appalling conditions, the Battle of Stalingrad came to symbolise the turn of the tide in the European war.

"Do you know what? We're modest; we've got it. And you can be quite certain that no one will ever be able to get us out now."
Adolf Hitler, Munich, November 1942

During the spring and summer of 1942 the German forces attacked the southern half of the front, from Voronezh to the Black Sea, and then advanced south towards the Caucasus and east towards the Volga. By September they had reached the industrial city of Stalingrad, the scene of some of the bitterest fighting. With few defences, the Russian 62nd Army on the right bank of the Volga, supported by artillery massed on the left bank, fought for every inch of territory. Losses were enormous on both sides. Hitler refused to withdraw any troops and, in November, the Red Army counter-attacked. It surrounded the German forces and forced them to surrender at the end of January 1943.

In the summer, the Germans attempted one further offensive. This, though, was relatively short-lived, the Germans having exhausted themselves in earlier campaigns. Conversely, the Red Army appeared to go from strength-to-strength: commanders began to display ability in the field; weaponry and equipment poured forth from the Allies and from factories relocated in the east; and on Soviet soil, partisans hampered German activities. Fortunately for the Soviet Union, the Battle of Stalingrad coincided with Montgomery's victory at El Alamein and Allied landings in North Africa. In June 1944, Allied forces landed in Normandy and established the "second front". As the Allies pushed into Germany from the west, the Soviets closed in from the east. From the autumn and throughout the winter of 1943/44, the Red Army recovered much of occupied Soviet territory and in April, advanced on central and eastern Europe. In January the Red Army invaded Germany on a broad front. Berlin fell on 2 May and on the 9th, Germany surrendered unconditionally.

Battle of Stalingrad 9–29 January 1943

- German advance
- German retreat
- German front line, 9 Jan
- German front line from 12 Jan
- ■ SIXTH ARMY German army
- ■ SIXTY-FOURTH ARMY Soviet army
- Soviet attack
- limit of Soviet artillery positions (approximately 200 artillery pieces per km of front line)
- bombardment

Factories moved East
- ○ grain elevator
- ○ oil storage area
- ○ Red October factory
- ○ Barrikady factory
- ○ tractor factory

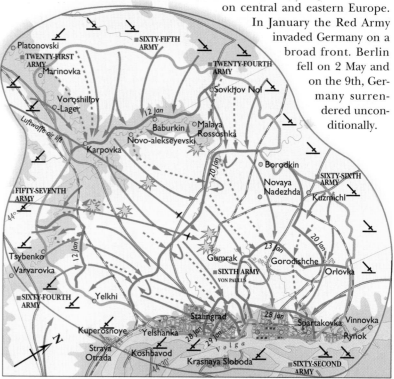

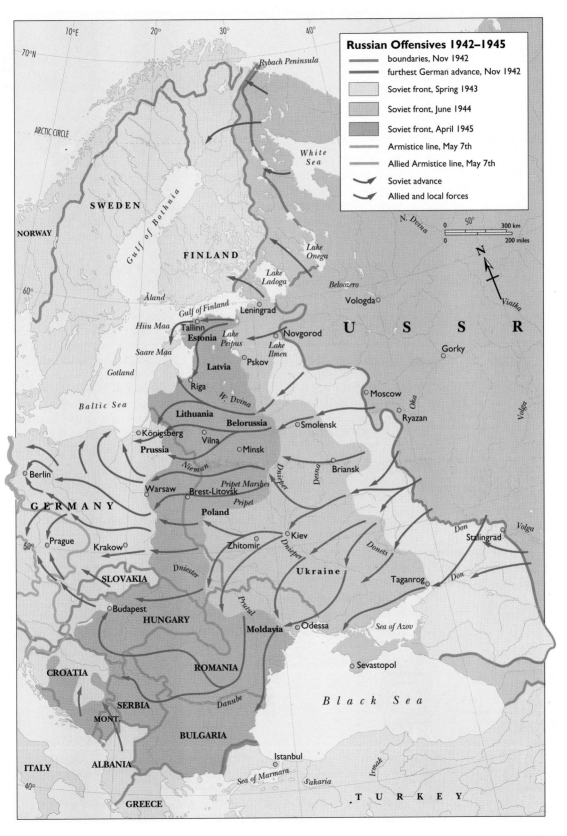

Russian Offensives 1942–1945

boundaries, Nov 1942
furthest German advance, Nov 1942
Soviet front, Spring 1943
Soviet front, June 1944
Soviet front, April 1945
Armistice line, May 7th
Allied Armistice line, May 7th
Soviet advance
Allied and local forces

Post-war Stalinism

As eastern and central Europe fell under the Soviet shadow, only Yugoslavia and Albania chose their own paths to Socialism.

"I will shake my little finger and there will be no more Tito!"
Joseph Stalin

At the end of the World War II the Soviet Union was recognised as a world power by dint of its contribution to the victory over Nazi Germany. As the Red Army advanced westwards it liberated Poland, Czechoslovakia, Hungary, Romania, Bulgaria and the eastern parts of Germany, thus extending the Soviet frontier by about 52,000 square miles to the west. This huge increase in territory added some 22 million people—from the Baltic States, Poland and Romania—to the Soviet population. The countries in the eastern and central European sphere of influence acted as a buffer zone for the Soviet Union, and in the immediate aftermath of war, provided some of the materials necessary for rebuilding Soviet industry. Multi-party systems were initially established in these states, but as relations with the West began to deteriorate, Communist Parties eventually took open control, establishing pro-Soviet governments between 1947 and 1948.

In March 1947, fearing the spread of Communism in Europe, President Truman offered to support any states faced by a communist takeover. This was the essence of the Truman Doctrine and the policy of containment. At the same time the Marshall Plan was introduced to help stimulate European recovery. Rejecting western economic aid, the Soviet Union brought pressure to bear on its Eastern European allies to do likewise, setting up Cominform (the Communist Information Bureau) as a successor to Comintern (the Communist International), with the task of co-ordinating the activities of the satellite states with those of the Soviet Union.

Mutual suspicion between the Soviet Union and the United States led to the creation of a bi-polar post-war world, based on the ideological cleavage between liberal capitalism and Marxism-Leninism. The peoples of eastern Europe were isolated from the West by the Iron Curtain.

The smiling face of Stalinism: Stalin at the All-Union Physical Culture Parade, July 1947. A master in the art of personality cults, Stalin emphasized the importance of his leadership in the victory over Germany, and played down the brilliant generalship of figures such as Zhukov.

Only Yugoslavia and Albania had the support of the majority of their peoples in establishing communist governments, and both diverged from the Soviet road to socialism in 1948 and 1960 respectively, awakening the western powers to the reality that the communist bloc was not as united as they had once believed.

Stalin, aware of this problem, realised that given the vote, none of the Eastern and Central European states would wish to remain within the Soviet sphere of influence. This meant that he could make few concessions to the particular needs of any individual satellite state.

Consequently, the peoples democracies of Hungary, Bulgaria, Albania, East Germany, Poland and Czechoslovakia were consolidated by means of a series of purges against both communist and religious leaders who were critical of Moscow and who put their national interests before those of the Soviet Union.

The new policy of "de-Stalinization", following Stalin's death in 1953, led to the widely held belief in eastern and central Europe that states might be allowed to move toward socialism along paths of their own choosing. Nevertheless, demonstrations and riots in Poland and East Germany in 1953 were ruthlessly suppressed, and the national rising in Hungary in 1956 was crushed by Soviet intervention.

Post-War Europe

Soviet pre-war territory

territory annexed by USSR 1939–40

former German & Czech territory annexed by USSR in 1945

states liberated by Soviet army

Soviet occupation zones in Austria and Germany

British, French and American occupation zones

annexed by Poland from Germany

'Iron Curtain' in 1948

Peoples resettled, evacuated or expelled

Germans   Finns   Russians   Russians forcibly repatriated

Czechs   Poles   Baltic peoples

Khrushchev and the Arms Race

The space race was the shop window of the arms race, and Khrushchev delighted in every Soviet extra-terrestrial achievement.

"... we felt pride in our country, our Party, our people and the victories they had achieved. We had transformed Russia into a highly developed country ..."
Nikita Khrushchev

The Khrushchev period was one of optimism—if not over-optimism—as demonstrated by Khrushchev's commitment to the "virgin lands" project which proved only partially successful. The best illustration of the optimism of the Khrushchev years was provided by the development of Soviet space and arms technology, and Khrushchev's realization of the importance of strategic nuclear weapons in foreign policy.

The space and arms races were to a large extent interlinked as the development of space rockets impinged upon that of missiles and atomic weapons, providing the world with a show case for Soviet technological and military achievement. After the war, German scientists were brought in to help Soviet research on atomic weaponry and rocketry. By 1949 the Soviets had developed their own atomic bomb.

On 4 October 1957, the USSR launched the world's first artificial earth satellite, Sputnik 1. A month later the dog, Laika, orbited the world in a space craft. The USSR had stolen a march on the United States. Soviet technological advancement came as an immense shock to the United States; from that moment the space race was on. By April 1961 Colonel Yuri Gagarin had

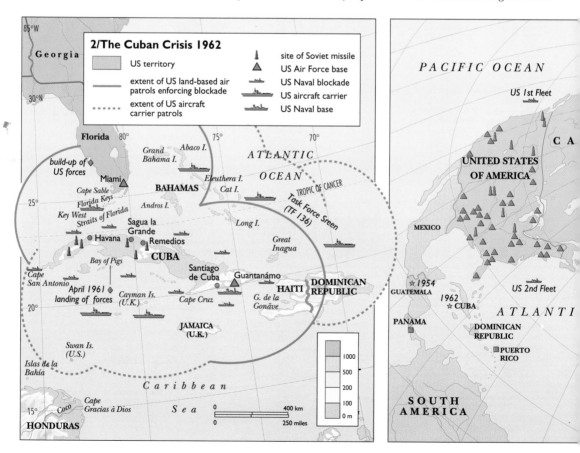

2/The Cuban Crisis 1962

- US territory
- extent of US land-based air patrols enforcing blockade
- extent of US aircraft carrier patrols
- site of Soviet missile
- US Air Force base
- US Naval blockade
- US aircraft carrier
- US Naval base

made the first manned space flight.

Khrushchev was enthusiastic about rocket technology, believing that it would provide the foundation for an inexpensive, secure defence for the Soviet Union. Sputnik 1 also provided him with a propaganda tool for foreign policy purposes; for several years he timed Soviet space launches to precede his major trips abroad.

Sputnik 1 had taken the West completely by surprise, suggesting that the Soviet ICBM programme must be further advanced than had been suspected, despite the fact that in reality it was well behind that of the United States. In the end this whole deception backfired as Khrushchev's "missile gap" led the United States into a defence build-up which pushed the gap into Washington's favour. Nevertheless, by the late 1960s, the Soviet military sector reached nuclear parity with the United States.

At the beginning of the 1960s, Khrushchev discovered that American Jupiter missiles, based in Turkey, could be launched against Kiev, Moscow and other major Soviet cities. This discovery, in October 1962, precipitated the greatest foreign policy crisis of Khrushchev's career. To counter the perceived imbalance of power he ordered Soviet medium-range rockets to be placed on Cuba, eliminating the intercontinental ballistic missile lead of the United States. Tension mounted rapidly; as President Kennedy demanded the removal of the missiles and threw a naval blockade around Cuba, the world seemed to teeter on the brink of war. Not until 28 October did the blustering Krushchev back down and order the removal of the missiles.

The ebullient Nikita Krushchev, First Secretary from 1953 to 1964.

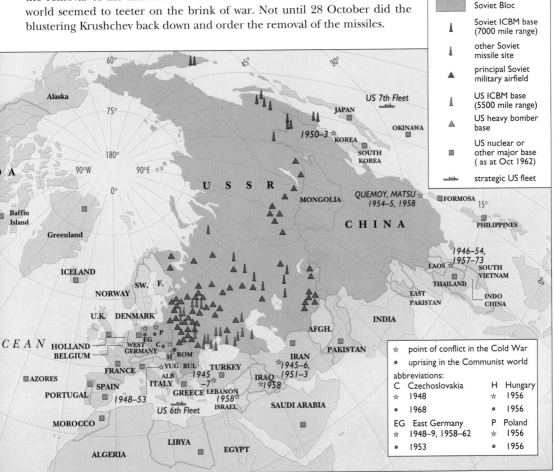

1/The Soviet and American Blocs, 1949–1959

Western Bloc

Soviet Bloc

Soviet ICBM base (7000 mile range)

other Soviet missile site

principal Soviet military airfield

US ICBM base (5500 mile range)

US heavy bomber base

US nuclear or other major base (as at Oct 1962)

strategic US fleet

☆ point of conflict in the Cold War

• uprising in the Communist world

abbreviations:

C	Czechoslovakia	H	Hungary
☆	1948	☆	1956
•	1968	•	1956
EG	East Germany	P	Poland
☆	1948–9, 1958–62	☆	1956
•	1953	•	1956

The Brezhnev Era

Under Brezhnev the dynamism of individual leadership gave way to stagnation under the rule of the Committee.

"Socialist states stand for strict respect for the sovereignty of all countries. We resolutely oppose interference in the affairs of any states and ... the violation of their sovereignty
Leonid Brezhnev, 12 November 1968.

By 1970 the Soviet Union had achieved military parity with the United States. In the field of foreign policy, Soviet influence had expanded in post-colonial Africa, Central America and Asia. In March 1969, tension with China almost flared into open conflict when clashes along the disputed frontier of the Ussuri river left some 30 Soviets and an unknown number of Chinese dead. The Soviet determination to dominate Eastern Europe was demonstrated by the invasion of Czechoslovakia in 1968.

Yet despite Soviet intervention in Czechoslovakia and growing tensions in the Middle East, East–West relations began to improve in the 1970s, as witnessed by the numerous summit meetings between Brezhnev and Presidents Nixon, Ford and Carter over détente and arms control. The symbolic Apollo–Soyuz joint space venture in 1979 well-illustrated this improvement in relations. Although the continuous round of Strategic Arms Limitation Talks (SALT) did not stop the arms race it did at least attempt to limit and control the number of nuclear weapons employed by both sides.

In December 1979 the Soviet Union embarked upon its biggest foreign policy undertaking since the Second World War by invading Afghanistan. With its troops bogged down in a costly war against the *Mujahidin* and facing an outburst of condemnation from both the western powers and the Islamic world, the Soviet government finally withdrew its troops in February 1989.

Intervention in Afghanistan, followed by the election of Ronald Reagan as US president in 1980, with his invective against the "evil empire", and the imposition of martial law in Poland in December 1981, led to a deterioration in East–West relations, which, with the shooting down of the Korean airliner (KAL 007) by a Soviet fighter-aircraft, resulted in what some commentators have termed as the "Second Cold War".

Disparities between Soviet military and civilian economies, and a growing technological gap between the USSR and western countries—highlighted by the implementation of Reagan's Strategic Defense Initiative (Star Wars)—meant that on Brezhnev's death in 1982 there was a growing need for change in Soviet foreign and domestic policies.

Leonid Brezhnev, General Secretary of the Party from 1966 onwards. Under Brezhnev high-ranking party officials were to achieve a standard of living the opulence of which could hardly be imagined by the average citizen. Nepotism and pluralism were rife, with Brezhnev himself holding no fewer than five high offices.

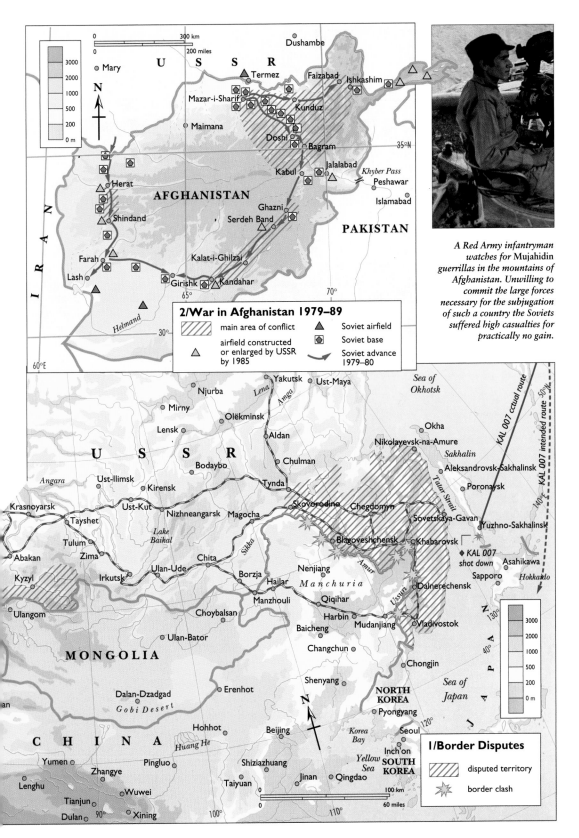

A Red Army infantryman watches for Mujahidin guerrillas in the mountains of Afghanistan. Unwilling to commit the large forces necessary for the subjugation of such a country the Soviets suffered high casualties for practically no gain.

Gorbachev – Nationalism and Europe

Attempts to liberalize the Soviet regime resulted in its flaws becoming ever more apparent.

"Characteristic of the national relations in our country are both the continual flourishing of the nations and nationalities and the fact that they are steadily drawing closer together on the basis of equality and fraternal cooperation.
The Programme of the CPSU, 1 March 1986

By 1991 Mikhail Gorbachev, though still lionized in the west, had become a discredited laughing-stock at home. In a survey asking "What does the Soviet Union give to its people?", 68 per cent answered "Shortages, queues and poverty".

In the 1980s the costs of the Afghan War and "Star Wars" rivalry were placing great strain on the Soviet economy. The USSR was spending between 11 and 15 per cent GNP on defence in comparison with between six and seven per cent in the United States. The concomitant cleavage between the military and civilian economies demonstrated a need for greater dialogue with the West and a series of reforms, known as *perestroika* and *glasnost*.

Perestroika was a combination of policies introduced by Mikhail Gorbachev at the XXVIIth Party Congress in 1986 which were implemented until the end of 1989. These policies brought changes to the economic and political system; for the first time since the end of NEP in 1929, elements of a market mechanism and multi-forms of ownership were introduced into the economy. *Glasnost* encouraged a greater openness to discussion, but without advocating the establishment of competing political parties. Attacks were made upon the bureaucratic state apparatus alongside campaigns against alcoholism and corruption. Environmental disasters, such as Chernobyl, were publicly revealed for the first time.

Political change was ushered in by the XXIXth Party Congress in July 1988 when Gorbachev proposed reforms of government structures, which resulted in an undermining of the Party's role and paved the way for a democratic Soviet Union when it was proposed that two-thirds of the seats of the Congress of People's Deputies should be open to competition on the basis of universal suffrage. During the 1989 elections a new phenomenon appeared on the Soviet scene, whereby important secretaries lost seats to little known individuals. One example among many was the rise to power of Boris Yeltsin as a reformer and "people's hero" to represent Moscow at the Congress of People's Deputies, despite official opposition.

By the end of 1989 the Party was losing its battle to retain its political pre-eminence. The Central Committee voted against the recognition of the Communist Party as the only institution of the political system, and by the summer of 1990 political parties were legalized. Soviet ideology had also failed to overcome the pressing problem of nationalism which now became a means of expressing growing discontent with central government. In March 1986 the Programme approved by the XXVIIth Party Congress emphasized the successful solution to the nationalities question in the Soviet Union, yet before the year was out the USSR was racked by inter-ethnic disputes and calls for independence. Influenced by events in Eastern Europe in 1989, nationalists in the Soviet Union struggled to take control from central authority and to defend their ethnic identities. The Soviet Union was visibly falling apart.

Russian nationalists demonstrate in front of the Lenin mausoleum on May Day, 1990.

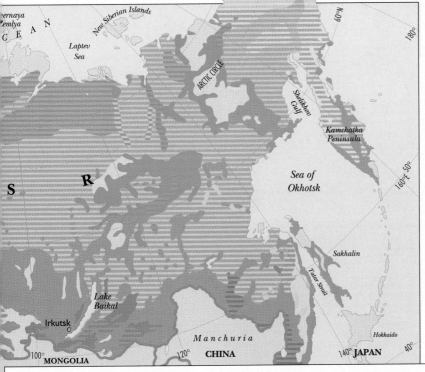

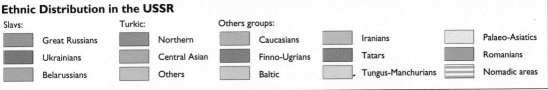

Ethnic Distribution in the USSR

Slavs:
- Great Russians
- Ukrainians
- Belarussians

Turkic:
- Northern
- Central Asian
- Others

Others groups:
- Caucasians
- Finno-Ugrians
- Baltic

- Iranians
- Tatars
- Tungus-Manchurians

- Palaeo-Asiatics
- Romanians
- Nomadic areas

USSR Dissolves

The growing weakness of the Soviet Union enabled its member states to push for independence.

"We are all like prisoners escaping from the same prison; we've made our escape but we're bound together by a single chain."
Yuri Scherbak, Ukrainian environmental minister

The political and social upheaval in the 15 successor states to the former USSR constitutes one of the most momentous events of the 20th century. Central to the dissolution of the Soviet Union was a dramatic rise of nationalism in the second half of the 1980s, voiced more openly as a result of *glasnost*. Consequently more national groups are recognised today—in Russia alone—than were recognised officially in the entire USSR before 1991.

In the last five years of Soviet power, millions of Soviet citizens, mainly from the non-Russian republics demonstrated their discontent against the ethnic, political, cultural and economic policies of the Soviet Union.

In the struggle for independence that followed, the relatively prosperous Baltic States of Lithuania, Latvia and Estonia were the first to rise against Soviet power, showing the way to other republics and ending the humility of occupation that they had endured since the Molotov–Ribbentrop pact. Their peaceful "national awakening" was symbolised by the much-televised human chain, in which two million people joined hands, linking the three capitals of Tallinn, Riga and Vilnius.

A landmark of capitalist endeavour in the centre of the communist world: the McDonald's fast food restaurant opened in Moscow, January 1990. With the disintegration of the communist bloc many western speculators have identified eastern Europe as a target for their entrepreneurship.

1 Adygeia
2 Karachaevo Cherkesia
3 Abkhazia
4 Adzharia
5 N. Ossetia
6 S. Ossetia
7 Chechya-Ingushetiya
8 Kalmykia
9 Dagestan
10 Nagorno-Karabakh
11 Nakhchevan

Elsewhere the road to independence had not been so peaceful because of deeper, simmering inter-ethnic disputes. For example, thousands of lives were lost, and people displaced as a result of the Armenian–Azerbaijani conflict over Nagorno–Karabakh as well as the ethnic tensions between Uzbekjs and Kazakhs, Georgians and Ossetians, Georgians and Abkhazians, and Tajiks and Russians, to say nothing of the fighting in the Dniester region of Moldova and the conflict in Chechnya.

With the break up of the USSR, 11 of the former Soviet republics formed a loose alliance termed the Commonwealth of Independent States (CIS). These new states have been faced with continuing economic crises. Ukraine, like Russia, has suffered the effects of rampant inflation, heralding the rise of black marketeers and local mafias. Against the background of corruption, racketeering, vice and growing social tensions, some Russian politicians, such as Aleksandr Rutskoy, refusing to accept the disintegration of the Soviet Union, have voiced their concerns over Russia's borders and general security. Concern has been expressed over the plight of the Russian ethnic minorities who remained in the Baltic States and who are now threatened with unemployment and the loss of citizenship rights in the light of local nationalism and strict language laws. Some Russian politicians advocate a return to "historic Russia", implying the revival of a pan-Slavic state made up of Russia, Belarus and Ukraine. Others, such as Vladimir Zhirinovsky, have even argued for reclaiming the "near abroad", the re-absorption of the Baltic states and other former Soviet territories.

The new Russian state emblem, adopted in 1991. In the late 1980s and early 1990s a number of political and social events became symbols of the collapse of communist rule: the demolition of the Berlin Wall, Leningrad's reversion to its former name of St. Petersburg, the toppling of the statue of Cheka's first, sadistic chief, Dzerzhinsky.

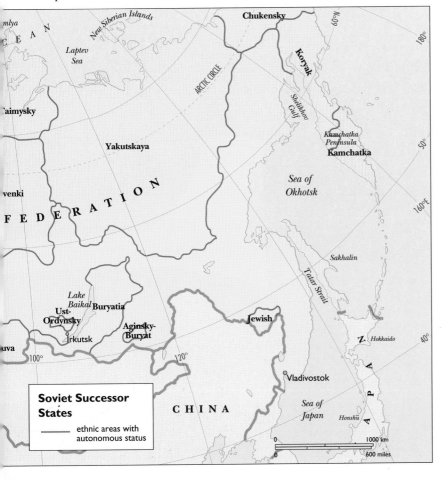

Soviet Successor States

——— ethnic areas with autonomous status

Khrushchev and the "Virgin Lands"

Following his rise to power Khrushchev immediately began improving the agricultural economy by an ambitious drive to develop the Soviet steppe.

"Communism ensures the continuous development of social production and rising labour productivity ... it equips man with the best and most powerful machines, greatly increases his power over nature and enables him to control its elemental forces to an ever greater extent."
Khrushchev's New Party Programme, 1961

As Prime Minister of the Ukraine under Stalin (1943–47), Khrushchev had had considerable experience in reorganising Soviet agriculture and amalgamating collective farms. When he became First Secretary in 1953, he advocated solving the problems of food supply by the cultivation of some 90 million acres of steppe land in north Kazakhstan, west Siberia and the south-eastern part of European Russia. Khrushchev wanted to develop these "virgin lands" as the major grain producing regions of the Soviet Union, and thus solve the Soviet Union's grain problem at a stroke. The Ukraine, the traditional "granary" of the Soviet Union, could then be used for growing maize for animal fodder. His ultimate aim was to overtake the United States in grain production by 1970.

The "virgin lands" programme employed thousands of tractors and combine harvesters, and was carried out by tens of thousands of young volunteers organised by the Komsomol, who moved from European Russia into

the steppe, where they lived in primitive conditions on state farms rapidly-built for the purpose.

By 1956, due to the initial fertility of the soil, the "virgin lands" produced three times as much grain in comparison with 1953. The experiment was hailed as a tremendous success.

The programme, however, had been rushed and fertilizers were not used to replace nutrients in the soil. Neither crop rotation nor irrigation methods were employed and the harvests began to decline. The situation was exacerbated by low rainfall, which led to droughts in the Urals in 1955, 1957 and 1958. Kazakhstan and the Volga were also hit by drought. Soil erosion began, aggravated by a series of storms between 1960 and 1965. Nearly half the "virgin lands" were turned into a dust bowl.

Although success had been only short-term, it nevertheless came at an important time; the "virgin lands" programme encouraged greater production on private plots; better incentives on the collective farms by offering higher procurement prices. In 1956, a drought occurred in the traditional grain-growing areas of the western Soviet Union which proved the importance of Khrushchev's scheme by saving the country from a serious shortage of grain. Nevertheless, by the 1963 the Soviet Union was forced to import grain from abroad.

Khrushchev admires the fruits of his labour in the "virgin lands". Though retaining the affection of many Russian people, in October 1964 Khrushchev was overthrown by the Presidium which denounced his "harebrained schemes, hasty conclusions, rash decisions and actions based on wishful thinking".

Further Reading

The following is a selective list which should be readily available to the general reader.

Acton, Edward, *Rethinking the Russian Revolution,* Edward Arnold, London 1990

Dawlisha, Karen and Parrott, Bruce, *Russia and the New States of Eurasia – The Politics of Upheaval,* Cambridge University Press, 1994

Diuk, Nadia and Karatnycky, Adrian, *New Nations Rising – The Fall of the Soviets and the Challenge of Independence,* John Wiley & Sons, Inc., New York 1993

Filzer, Donald, *The Khrushchev Era – De-Stalinisation and the Limits of Reform in The USSR, 1953–1964,* Macmillan, London 1993

Hingley, Ronald, *Russia, a Concise History,* Thames and Hudson, London 1991

Höetzsch, Otto, *The Evolution of Russia,* Thames and Hudson, London 1966

Hosking, Geoffrey, *A History of the Soviet Union,* Fontana, London 1985.

McAuley, Mary, *Soviet Politics, 1917–1991,* Oxford University Press, 1992

McCauley, Martin, *The Soviet Union since 1917,* Longmen, London 1981

Merridale, Catherine and Ward, Chris (eds.) *The Historical Perspective – Perestroika,* Edward Arnold, London 1991

Miller, John, *Mikhail Gorbachev and the End of Soviet Power,* Macmillan, London 1993

Milner – Gulland, Robin and Dejevsky, Nikolai, *Atlas of Russia and the Soviet Union,* Phaidon Press, Oxford, 1989

Riasanovsky. N.V., *A History of Russia,* Oxford 1984

Service, Robert, *The Russian Revolution, 1900–1927,* Macmillan, London 1991

Stokes, Gale (ed.), *From Stalinism to Pluralism,* Oxford University Press, 1991

Stone, Norman, *The Russian Chronicles,* Random Century, London, 1990

Tompson, William, *Khrushchev – A Political Life,* Macmillan, London 1995

White, Steven, *After Gorbachev,* Cambridge University Press, 1993

Wood, Alan, *The Origins of the Russian Revolution, 1861–1917,* Methuen, London 1992

Index

PLACES

Acknowledgements

Picture Credits

Arxiu Mas: 22

Bridgeman Art Library, London:
  35 Novosti; 52 Lauros-Giraudon; 61
  Novosti; 48 Private Collection; 93
  Musee D'Orsay, Paris; 99 Tretyakof
  Gallery, Moscow

C. M. Dixon: 19

Codex: 47, 90, 120

David King Collection: 75br, 104, 108, 110, 112,
  127, 134-135

E. T. Archive: 66

Jean-Loup Charmet, Paris: 65, 79cl, 83

John Massey Stewart: 24, 40, 72t, 72b, 94

Lenin Museum, Moscow: 107

Mansell Collection: 59b

Michael Holford: 42-43, 58br, 58cl, 86

Novosti, London: 26br, 27, 33, 38, 49, 50, 62,
  74t, 74b, 78, 82, 84, 95, 97, 98-99, 100, 109,
  130, 132, 133

Peter Clayton: 18

Popperfoto: 113

Robert Harding Picture Library: 26cl

Society for Cooperation in Russian and Soviet
  Studies, London: 17, 30, 32, 34, 43, 59t, 79b,
  106, 114, 115, 116, 118, 119, 124, 128, 129,
  131

The Fotomas Index: 57, 76

FOR SWANSTON PUBLISHING LIMITED

Concept:
Malcolm Swanston

Editorial:
Stephen Haddelsey
Caroline Lucas
Chris Schüler

Editorial Assistance:
Rhonda Carrier

Illustration:
Ralph Orme

Cartography:
Andrea Fairbrass
Peter Gamble
Elsa Gibert
Elizabeth Hudson
Thekla Kempelmann
Isabelle Lewis
David McCutcheon
Kevin Panton
Peter Smith
Malcolm Swanston
Simon Yeomans

Additional Cartography:
Advanced Illustration,
Cheshire

Index:
Jean Cox
Barry Haslam

Typesetting:
Jeanne Radford

Picture Research:
Caroline Lucas
Charlotte Taylor

Production:
Barry Haslam

Separations:
Central Systems,
Nottingham.
Quay Graphics,
Nottingham.